I0471013

Exploration and Exploitation

of the 3 cm to 3 mm Wavelength Region

By Doc Ewen (Harold I. Ewen, Ph.D.)

Originally published 1970 as part of
Advances in Microwaves, Volume 5
Edited by Leo Young
Published by Academic Press, Inc.

Reprinted and published by
Ewen Prime Company
David K. Ewen, M.Ed.

With permission by
Doc Ewen (Harold I. Ewen, Ph.D.)

ISBN-13: 978-1491017296

ISBN-10: 1491017295

Preface

The fifth volume of *Advances in Microwaves* contains three chapters that range in their coverage from low microwave frequencies used to accelerate elementary particles, through cm and mm waves for exploring atmospheric phenomena, and on to the microwave demodulation of light.

The chapter on high-speed photodetectors for recovering microwave signals modulated on a laser carrier is our first successful collaboration between authors working for different companies. L. K. Anderson, M. DiDomenico, Jr., and M. B. Fisher are familiar with both microwaves and lasers. When large bandwidths are to be transmitted on light beams, the signal must be (1) modulated onto the laser carrier frequency, (2) transmitted, and (3) demodulated. The chapter deals with the third topic. (The second topic was considered in G. Goubau's chapter in Volume 3 and A. E. Karbowiak's chapter in Volume 1; the first topic may be the subject of a future volume.)

It is a pleasure to include a chapter by H. I. Ewen. He has been a pioneer in microwave radiometric measurements through and of the atmosphere. In the past, the frequency decade 10 to 100 GHz has been used to probe the atmosphere and has yielded much meteorological information. This frequency band has long held out promise for microwave communications, a promise that seems to be on the point of being fulfilled via satellites in space.

We have included another contribution from abroad. French author Y. Garault writes on microwave hybrid modes, which are used to deflect and separate high-energy particles in the linear accelerators at CERN in Europe, and at Brookhaven and Stanford in America. We wish to acknowledge the help and advice received from G. A. Loew in preparing this chapter. The reader is also referred to the chapter on the Stanford linear accelerator in Volume 1.

This volume could not have been assembled without use of the facilities at Stanford Research Institute. We are also grateful to Miss Dianna Bremer for her unfailing help in many ways.

LEO YOUNG

Exploration and Exploitation of the 3 cm to 3 mm Wavelength Region

Harold I. Ewen

EWEN KNIGHT CORPORATION
EAST NATICK, MASSACHUSETTS

I.   INTRODUCTION

Historically, the use of the millimeter portion of the spectrum has undergone cyclic periods of interest. Now increased attention has been directed to this wavelength region, spurred in large part by the worldwide explosion in communication needs. There are many inducements to consider this portion of the spectrum from a communication standpoint. Relatively high antenna gain is achieved with modest aperture diameter; broad channel capability permits high information capacity, and the total available bandwidth, even within the restricted atmospheric windows, far exceeds the entire radio spectrum below 10 GHz (3 cm wavelength). Exploitation of this wavelength region for communication has, in large part, been paced by the need for reliable millimeter power generating devices and low-noise receiving systems. The required technological advancements in these areas appear imminent.

Though the prime interest, as in the past, has been the need to alleviate congestion in the microwave communication bands, the latest resurgence of interest in millimeter waves has been aided by a passive, but not silent, partner. Exploration of this portion of the spectrum has been forging ahead at an accelerating pace through the application of passive radiometric measurement techniques. Some of these investigations are concentrated in the available "atmospheric windows" to establish their future potential for earth-space communication links. Several significant investigations, however, are being directed to those portions of the spectrum where the level of atmospheric opacity is too great to be useful for communication. Radiometric sensing of the electromagnetic emission of the atmosphere in these portions of the spectrum is providing a new and powerful tool for the investigation of atmospheric structure and the associated physical processes. Today, we are at the dawn of the new science of microwave meteorology. We can expect many startling discoveries beginning in the decade of the 70s, as microwave and millimeter radiometric sensors contribute to the challenge of global weather prediction.

The new field of microwave meteorology was spawned by the young science of radio astronomy which has produced so many startling discoveries concerning our galaxy and the universe. From the earliest experiments performed in the HF and VHF bands, the radio astronomer's spectrum of interest has progressed toward the millimeter wavelength region, paralleling the move of communication systems to higher frequencies, with the upward step for each paced by advancements in instrument technology. Exploration and exploitation of the higher frequencies has historically favored the radio astronomer since the passive receiving devices needed for radio telescopes frequently become available before the power generating devices needed for communication systems are developed. Each upward step in the spectrum has led to unanticipated discoveries. The significance of these discoveries has, on occasion, suggested the exclusion of communication systems from certain portions of the microwave and millimeter spectrum. Radio astronomers and communicators share those portions of the spectrum frequently referred to as the "atmospheric windows" where electromagnetic radiation passes through the atmosphere with least attenuation. These windows, which are centered near wavelengths of 8 and 3 mm, are open during clear weather conditions, partially closed by heavy water laden clouds, and are essentially closed during occasional periods of heavy rain. The attenuation and noise characteristics of the atmosphere in these windows are of prime concern to both the radio astronomer and communicator. The astronomer must understand the propagation characteristics of the atmospheric medium in order to delete its contaminating effects from the analysis of the very faint signals received from space. The communicator must know how the atmospheric medium effects signal fading, angle modulation, and correlation bandwidth in order to determine the optimum system design. Several significant communication research efforts in this area, today, are based on techniques developed in the field of radio astronomy. An obvious reciprocal benefit will be knowledge gained by the young science of microwave meteorology.

The microwave radiometer is the common denominator in the exploration of the 3 cm to 3 mm wavelength region. Invented by Dicke [1] less than three decades ago, embellished and exploited by radio astronomy, its use is rapidly spreading to a diversity of scientific research and engineering disciplines and applications in the explosive pioneering exploration of the millimeter wavelength region. It will be helpful in discussing these applications if we first review certain radiometric fundamentals associated with this portion of the frequency spectrum.

II. MICROWAVE RADIOMETRY

A microwave radiometric sensor is a device for the detection of electromagnetic energy which is noise-like in character. The spatial as well as spectral characteristics of observed energy sources determine the performance requirements imposed on the functional subsystems of the sensor. These subsystems include an antenna, receiver, and output indicator. Natural or non-man-made sources of radiation may be either spatially discrete or extended. In the frequency domain, these sources may be either broadband or of the resonant line type. Sensor design and performance characteristics are primarily determined by the extent to which spatial and frequency parameters characterize the radio noise source of interest to the observer.

A microwave radiometric sensor is frequently referred to as a temperature measuring device, since the output indicator is calibrated in degrees Kelvin. The reason why microwave radiometers are calibrated in temperature units and the modes of operation that are most frequently used are described in the sections immediately following.

A. TEMPERATURE CALIBRATION OF THE OUTPUT INDICATOR

The physical reasoning in support of calibrating the output indicator of a microwave radiometer in degrees Kelvin can be derived from thermodynamic considerations and certain well-known properties of an antenna.

The amount of energy absorbed by an antenna and presented at the input terminals of the receiver depends upon the orientation of the antenna, the polarization of the wave, and the impedance match of the receiving system. Since all antennas are polarized, regardless of design, the maximum amount of energy accepted by an antenna, from a randomly polarized wave, is one-half of the total energy content of the wave. If we assume that an antenna is perfectly matched and that the incoming wave is randomly polarized with a power flux density S, then the absorbed power P_A is given by the expression

$$P_A = \tfrac{1}{2}SA \qquad (1)$$

where A is the effective antenna aperture area.

In Eq. (1), the flux density S of the radiation is assumed to be from a source of small angular size and is measured by the flow of energy from the source through unit area in the wave front at the observing point. If energy dE in the frequency range $d\nu$ flows through area dA in time dt (where dt is long compared to the period of one cycle of the radiation), then the flux density S is given by the expression

$$S = \frac{dE}{dA\,d\nu\,dt} \qquad (2)$$

which has the dimensions of power per unit area per unit bandwidth.

Now consider a transmission line, one end of which is terminated with a matched load and the other end of which feeds an antenna in an absorbing medium. If we were to replace the antenna by its equivalent two-terminal network and assume that it is a purely resistive impedance and equivalent to the load impedance, then a transmission line terminated in a matched antenna may be treated in a manner similar to a transmission line terminated with a resistive load, as shown in Fig. 1. If the extent of

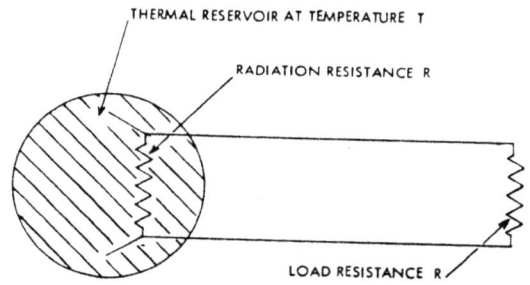

FIG. 1. Equivalent circuit of an antenna immersed in an absorbing medium at temperature T. In equilibrium, the temperature of the load resistance is the same as the temperature of the absorbing medium.

the absorbing medium is sufficient to completely absorb all radiation from the antenna, the medium and the matched termination must then be at the same temperature T.

From Johnson noise power considerations, the termination will radiate a power $kT\,d\nu$ to the antenna. If the antenna, in turn, did not accept $kT\,d\nu$ of radiation from the medium and transfer this power to the load, there would be a net transfer of thermal energy from one region to another at the same temperature without application of work, in violation of the second law of thermodynamics. This would indicate that in the micro-wave and millimeter portion of the spectrum, the power delivered to the receiving system input by an antenna immersed in an absorbing medium at temperature T is independent of the frequency of observation.

This conclusion can also be reached (see Fig. 2) by noting that the medium appears as a blackbody to the radiation resistance of the antenna, i.e., it absorbs all incident radiation and its radiation brightness β in the frequency interval $d\nu$ in accord with Planck's law is

$$\beta \, d\nu = \frac{2h\nu^3}{c^2} \frac{d\nu}{[\exp(-h\nu/kT) - 1]} \tag{3}$$

where

$$h = \text{Planck's constant}$$

$$k = \text{Boltzmann constant}$$

$$c = \text{velocity of light}$$

and the brightness β is the power per unit area per unit solid angle per unit bandwidth.

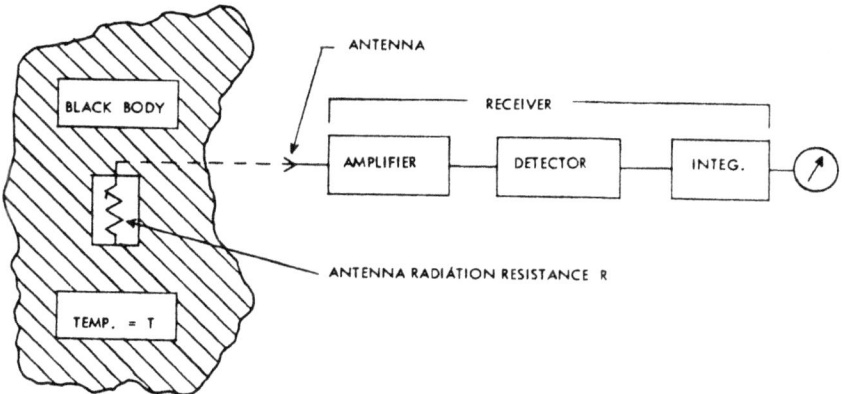

FIG. 2. Simplified block diagram of an antenna and receiver. When the antenna is immersed in a blackbody at temperature T, the receiver input is equivalent to a resistive load R immersed in a thermal bath at temperature T.

From the definition of flux density S, it is evident that

$$S = \int \beta \, d\Omega \tag{4}$$

where $d\Omega$ is the solid angle increment. This definition of flux density holds for any source of radiation over all solid angles.

Equation (3) indicates that the power received from a blackbody, measured at the input terminals of the receiving system, is frequency dependent. Hence, the equivalence of blackbody radiation and Johnson noise power appears inconsistent. The answer to this paradox lies in the characteristic frequency response of any antenna system and the frequency characteristic of blackbody brightness in the millimeter and microwave portions of the spectrum. This can be most easily seen by recalling that the power absorbed by an antenna, from a randomly polarized source when operating in the frequency range $d\nu$, is

$$P_R = \frac{1}{2} A(\theta, \phi) \beta \, d\nu \, d\Omega \qquad (5)$$

where θ and ϕ describe the direction of the incoming wave and $A(\theta, \phi)$ is the antenna aperture receiving cross section in that direction. Hence, for an extended source

$$P_R = \frac{1}{2} \iint A(\theta, \phi) \, d\Omega \, \beta \, d\nu \qquad (6)$$

If the extended source radiation is a blackbody, the power received by the antenna would then be expressed in the form

$$P_R = \frac{1}{2} \iint A(\theta, \phi) d\Omega \frac{2h\nu^3}{c^2} \frac{d\nu}{[\exp(h\nu/kT) - 1]} \qquad (7)$$

In that portion of frequency spectrum where the energy of the photon $h\nu$ is much less than the random thermal energy per degree of freedom kT at temperature T, the expression for blackbody brightness [Eq. (3)] reduces to the simplified expression

$$\beta \, d\nu \approx \frac{2kT}{\lambda^2} d\nu \quad \text{(Rayleigh–Jeans)} \qquad (8)$$

where λ is the wavelength of observation. Hence, in this portion of the spectrum Eq. (7) reduces to

$$P_R = \frac{kT}{\lambda^2} d\nu \int A(\theta, \phi) \, d\Omega \qquad (9)$$

Recalling that the average effective cross section for any antenna immersed in a source of uniform brightness may be expressed in the form

$$\bar{A} = \frac{1}{4\pi} \int A(\theta, \phi) \, d\Omega = \frac{\lambda^2}{4\pi} \qquad (10)$$

or

$$\int A(\theta, \phi) \, d\Omega = \lambda^2$$

we arrive at the conclusion that the power received by an antenna immersed in a blackbody at temperature T is frequency independent and equivalent to the Johnson noise power $kT\,d\nu$. As a consequence, the power received by a microwave or millimeter radiometer is conventionally described in terms of equivalent temperature units.

The transition region in the frequency spectrum at which the energy of the photon is comparable to the random thermal energy per degree of freedom is, of course, temperature dependent. Approximate values are shown in Table I. It is apparent from Table I that microwave and millimeter wavelength measurements of the earth terrain and atmosphere (ambient 290°K) at frequencies below 300 GHz ($\lambda = 1$ mm) fall well within that region of the spectrum where $h\nu$ is less than kT.

Table I

TEMPERATURES AND CORRESPONDING WAVELENGTHS
AT WHICH THE ENERGY OF THE PHOTON $h\nu$ IS EQUAL TO kT

Temperature (°K)	Wavelength (μm)	Wavelength (mm)	Frequency (GHz)
300	70	0.07	4300
77	200	0.2	1500
20	900	0.9	333
4	6000	6.0	50
1.5	10,000	10.0	30

In summary, the power received by an antenna immersed in a blackbody at a temperature T is frequency independent and equivalent to the Johnson noise power that would be radiated by an antenna if terminated in a matched resistive load at the same temperature T. These two fundamental sources of noise power are equivalent at microwave frequencies due to the inverse wavelength squared dependence of blackbody brightness, which is offset by the wavelength squared dependence of the antenna cross section. Hence, the noise power per unit cycle received by an antenna and presented at its output terminals is directly proportional to the effective blackbody temperature which characterizes the source or sources in which the antenna pattern is immersed. The proportionality factor is Boltzmann's constant k.

Since most natural sources are not blackbodies, their "signal temperature," measured by a radiometric sensor, refers to the power level that would be received from a blackbody at a temperature which would provide an equivalent power level at the output terminals of the antenna.

This temperature concept is useful in describing the functions of the antenna and receiver in a microwave radiometric sensor. The antenna extracts noise power from the radiation incident on its aperture and presents a noise power at its output terminals which can be described in terms of an effective blackbody temperature. This noise power represents a composite of the desired signal power and undesired noise power from other sources, since a practical antenna always looks to some degree in undesired directions: or the signal source may be immersed in a background noise field.

If the effective temperature of the composite noise power presented at the output terminals of the antenna is T_A and that portion associated with useful signal power is T_S, then $T_A = T_S + \sum T_i$, where $\sum T_i$ represents a summation of effective noise temperatures from the undesired sources of noise power observed by the antenna. The signal-to-noise ratio at the output terminals of the antenna is then $T_S / \sum T_i$. The prime function of the receiver is to amplify and detect the input signal which is characterized by the composite temperature T_A. All processes of receiver amplification add noise to the received signal. This added noise is frequently referred to as the internal receiver noise which can be described by an effective temperature T_R referred to the input terminals of the receiver. The ratio of antenna temperature to receiver noise temperature, at the interface between the antenna and receiver, is then $(T_S + \sum T_i) / T_R$. Note that the unwanted noise power $\sum T_i$, received by the antenna and presented at the input terminals of the receiver, cannot be differentiated from the desired signal temperature T_S through amplification alone. Spatial differentiation between T_S and $\sum T_i$ may be obtained by scanning of the antenna beam, if the source of signal temperature T_S is spatially discrete and the sources contributing to $\sum T_i$ are spatially extended. Similarly, the separation of T_S from $\sum T_i$ may be obtained in the frequency domain by the receiver, if either the source of signal temperature or background noise temperature exhibit markedly different frequency characteristics, such as a resonant line superimposed on a broadband continuum. In this case, the receiver can be scanned in the frequency domain to separate the signal temperature from the temperatures contributed by broadband background sources.

The determination of the equivalent noise temperature of a receiving system is related to the method of noise figure measurement. The noise figure of a receiving system or network is defined as the signal-to-noise ratio at the input, divided by the signal-to-noise ratio at the output, when the receiver or network is terminated in a matched load at a temperature T_0 of $290°K$. A simplified equivalent receiver network, shown in Fig. 3, consists of a network with input terminals shunted by a resistor R and output terminals connected to a meter indicator. From the definition of noise figure F, the noise figure of the network shown in Fig. 3 is given by the expression

$$F = \frac{Gk T_0 \, d\nu + GN}{Gk T_0 \, d\nu} \qquad (11)$$

or

$$F = 1 + \frac{N}{k T_0 \, d\nu} \qquad (12)$$

If we now define the system noise temperature T_R by the expression

$$N = k T_R \, d\nu \qquad (13)$$

we see that the relationship between system noise temperature and noise figure is

$$T_R = (F - 1) T_0 \qquad (14)$$

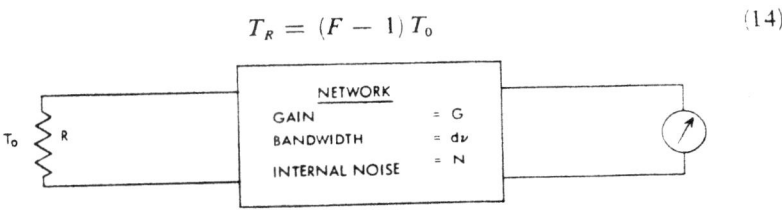

FIG. 3. Input and output signal and noise relationships of a four-terminal network with the input terminated in a resistive load which is immersed in a thermal bath at temperature T.

B. RECEIVER FUNCTIONS AND TECHNIQUES

The prime function of the receiver in a microwave radiometric sensor is to provide a measure of the antenna temperature.

As previously noted, the antenna temperature T_A is, by definition, the temperature to which the radiation resistance of the antenna must be raised in order to produce the same noise power as that contributed by the various sources observed by the antenna. It is also the brightness temperature of a blackbody which, if it completely surrounded the antenna, would provide the same noise power at the receiver input. To describe the method by which the receiver measures the antenna temperature, we will replace the antenna with an equivalent resistive load at the receiver input. If the antenna temperature were T_A, we would obtain the same noise power input to the receiver by placing the resistive load in a thermal bath at a temperature T_A.

The need for signal amplification becomes readily apparent when one notes that the average noise power per unit bandwidth produced by a resistor at an ambient temperature (290°K) is of the order of 10^{-20} watt. Typical detectors require a drive power of at least 10^{-9} watt. The required input signal amplification must, of course, be increased if temperature changes less than 290°K are to be detected and recorded. The receiver must therefore be able to sense a low level change in noise power at its input and provide sufficient stable amplification to drive the output indicator system. Amplification stability is a prime requisite since the receiver must provide a consistent output response for the same input power change. The relatively poor gain stability of present receiving systems is overcome by the use of an input switch or modulator, to be discussed later.

1. *Sensitivity*

The noise power output of a resistive termination is associated with the thermal agitation of electrons within the resistive conductor which produce electronic collisions. As the thermal temperature of the resistor is increased, the thermal agitation increases; and the number of collisions per unit time increases. The resultant noise power output per unit cycle is directly proportional to the absolute temperature of the resistor. As indicated previously, the proportionality factor is Boltzmann's constant k. In this sense, a radio measurement of the thermal temperature of the input resistor may be described as a measurement of the electron collision frequency within the resistor. Since the collisions are random, the number per second will vary; however, the mean of an infinite number of one-second samples will lead to an exact value for the collision frequency. From statistical theory, the probable error in the measurement of a quantity of this type is inversely proportional to the square root of the number of measurements which are made. If the number is infinite, the exact value is determined. If we now measure the electronic collisions within a resistor, using an amplifier of finite bandwidth $\Delta \nu$, the number of independent collisions per second which can be counted is equivalent to the receiver bandwidth. Hence, the error in determining the mean value of the noise temperature (which is proportional to the collision frequency) will be inversely proportional to the square root of the receiver bandwidth. If the averaging process is extended over τ seconds rather than one second, there will be, on the average, $\tau \Delta \nu$ independent collisions in each interval of seconds, therefore

$$\frac{\Delta T_R}{T_R} \approx \frac{1}{\sqrt{\tau \, \Delta \nu}} \tag{15}$$

In most radiometric applications, the magnitude of the signal temperature is negligible when compared with the "receiver noise temperature T_R" which describes the noise power added to the received signal by the various circuits within the receiver.

Simplified functional block diagrams of the most commonly used microwave and millimeter radiometric receiving systems are shown in Fig. 4. The figure depicts the genealogical growth of each receiver from

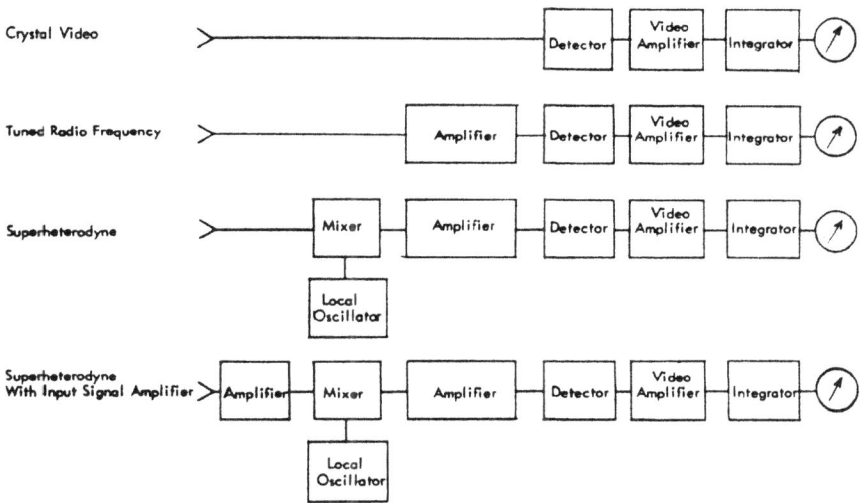

FIG. 4. Simplified block diagrams of commonly used radio receivers.

the one preceding. The crystal video receiver is usually the first to be used in a new portion of the frequency spectrum where input signal frequency components required for the other modes of operation are not available. The superheterodyne is the "workhorse" among receivers. The input circuit of the superheterodyne is a "mixer" in which the signal frequency is heterodyned with the local oscillator frequency. The difference or intermediate frequency between the signal and local oscillator is amplified by a tuned intermediate frequency amplifier, referred to as the IF amplifier. The addition of a low noise amplifier forward of the mixer in a superheterodyne mode will establish the receiving system noise temperature by providing adequate gain to overcome the conversion loss of the mixer. The most sensitive broadband radiometers operate in the TRF mode, where amplification is provided by the cascading of broadband low noise amplifiers, typically, traveling wavetubes and, more recently, tunnel diode amplifiers. Today, completely solid-state TRF receivers, using tunnel diode amplifiers, provide nominal system noise temperatures of 1000°K and instantaneous predetection bandwidths of 10 to 15% of their operating frequency, at frequencies up to 20 GHz.

The sensitivity of a radiometric system, i.e., the minimum detectable signal, is determined by the amplitude of the fluctuations present at the output indicator in the absence of a signal. These fluctuations are attributable to two sources:

(1) The statistical fluctuations in a noise waveform as described by Eq. (15).

(2) Spurious gain fluctuations associated with the receiving network. The amplitude of output fluctuations due to the first source can, in principle, be reduced to any desired degree by reducing the postdetection bandwidth (increasing the integration time). In practice, however, the longest usable integration time is limited by the time available for observation of the "signal."

2. Gain Variations and the Dicke Mode

The second source of fluctuations which occur at the receiver output are attributable to receiver gain instabilities. Their significance can be readily grasped by the following example. If we introduce values of $T_R = 1000°K$, $\Delta\nu = 2 \times 10^9\,Hz$, and $\tau = 1\,sec$ in Eq. (15), we obtain an rms value for the amplitude of statistical noise fluctuations at the receiver output of the order of 0.03°K. This would be the case if the receiver were absolutely gain stable. Unfortunately, the best receivers, regardless of type or frequency of operation, exhibit gain instabilities of the order of 1% during a time period comparable to that required for a noise measurement. As a consequence, a receiver with the performance characteristics described above would provide an output fluctuation of 10°K if the gain changed by 1%. The noise measurement sensitivity of

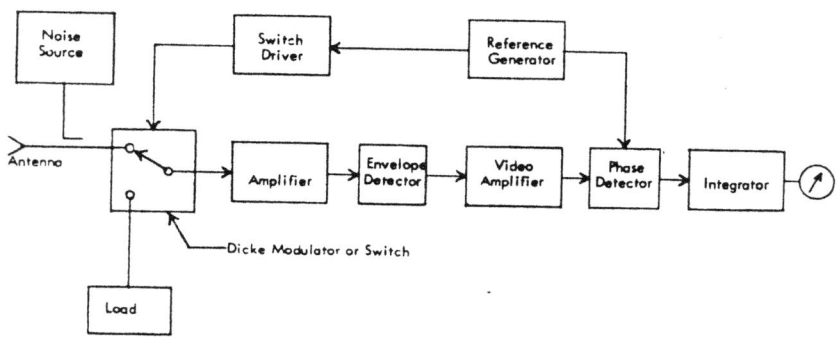

FIG. 5. Simplified block diagram of the Dicke radiometer. A switch or modulator is introduced between the antenna output and receiver input.

the system would then be determined by the effect of gain variations rather than by the level of statistical noise fluctuations.

The answer to this dilemma was provided by Dicke [1] in the form of a single pole, double throw switch placed at the input of the receiver, as shown in Fig. 5. One of the input ports of the switch is connected to the antenna output terminal, the other to a resistive load held at a constant temperature T_C. The switch is driven sequentially in a square wave fashion at a frequency considerably higher (typically 30 to 1000 Hz) than that at which a substantial receiver gain variation occurs. With the switch in operation, a signal at the switching or modulation frequency is presented at the input terminals of the receiver with an amplitude proportional to the temperature difference $T_A - T_C$. Because of the rapid switching rate, any receiver gain variation will operate equally on $T_A + T_R$ during one-half of the switching cycle and on $T_C + T_R$ during the other half, with the result that it operates only on the difference $T_A - T_C$. If for example, the difference $T_A - T_C$ were $1°K$, the effect of a 1% receiver gain variation referred to the output indicator system would be $0.01°K$.

In the example given above, the introduction of the switch provided a marked improvement in the noise measurement capability of a typical receiver by eliminating the effect of receiver gain variations operating on the receiver noise temperature. The gain variations, however, continue to operate on the temperature difference presented at the two input ports of the switch. This was not an important consideration in early radiometers which had relatively high noise temperatures and narrow bandwidths, leading to sensitivities of the order of a few degrees Kelvin. Present-day broadband radiometric receiving systems, however, have potential sensitivities of the order of $0.05°K$ rms for postdetection time constants of 1 second or less at frequencies up to and including 20 GHz. In this case, the effect of receiver gain variations, operating on an RF input unbalance (large temperature difference between input signal and comparison ports), is of far greater concern. Several techniques for reducing the RF input temperature unbalance are in common use. These include addition of noise to the signal port of the radiometer, use of a low temperature comparison source, and introduction of gain modulation.

Addition of noise in the signal transmission line is frequently reserved for applications in which the system noise temperature is relatively high, i.e., such that the added noise represents a small percentage increase in the overall system noise level. Radiometers with maser or low noise parametric input amplifiers normally use a low temperature comparison source such as a resistive load immersed in a liquid helium bath.

The technique of "gain modulation," introduced approximately one decade ago, involves the adjustment of receiver gain in synchronism with the switch or modulation frequency to provide an equivalent level of noise at the input to the envelope detector during both portions of the switch cycle. This technique provides a convenient adjustment of the effective temperature level at either input port without adding noise to the receiver or changing the temperature of the comparison noise source. The gain modulator technique, however, is sensitive to changes in system noise figure and must be used with caution.

3. Temperature Calibration

The detection of a signal noise source and its measurement in absolute temperature units represents a prime objective in several radiometric receiving system applications. In addition to sensitivity established by the noise characteristics of the detector and, in large part, influenced by the gain stability of the overall receiving system, measurements of this type require:

(1) Knowledge of the output indicator zero level in absolute temperature units.

(2) Calibration of the output indicator deflection in absolute temperature units.

The output indicator zero level corresponds to the condition of RF temperature balance between the input signal and comparison ports (switch ports) of the receiver. Under this condition, the comparison source temperature referenced to the comparison input port provides the indicator "zero level" for the indicator scale. Calibration of the output indicator deflection requires knowledge of the detector law and system gain. Of these two requirements, the indicator zero level is by far the more difficult to achieve. Knowledge of the detector law can easily be obtained through laboratory measurement. System gain can be established at any time during a measurement program by introducing a constant and fixed level of noise at the radiometer input. The noise level of this gain calibration noise source need not be known precisely. It is far more important that it remain constant and that it be used to establish the level of receiver gain during laboratory calibration of the radiometer response in equivalent temperature units. This noise source is usually included as an integral part of a radiometer and is referred to as the "calibration or internal noise source."

Calibration of the output indicator requires the introduction of a precisely known temperature change at the input signal port of the radiometer. This measurement is usually performed under carefully controlled laboratory conditions. This temperature change is frequently generated by the sequential introduction of two very precisely known noise temperature sources at the signal input port of the radiometer. Calibration of the internal noise source in equivalent temperature units is automatically obtained as a by-product of this laboratory calibration procedure.

From the foregoing, it is apparent that one internal fixed calibration noise source, combined with knowledge of the detector law characteristic and predetection attenuator values, provides all of the information required for the precise calibration of the output indicator reading in equivalent temperature units. The internal calibration noise source level should be approximately two orders of magnitude greater than the amplitude of the peak-to-peak fluctuation level at the output indicator for the nominal value of postdetection integration time constant which will be used during the measurement program. This allows opportunity to establish the full-scale output indicator deflection level to an accuracy of at least 1%. When measuring the amplitude of low level signal temperatures, the introduction of the indicator calibration signal will normally require an output indicator scale change. This is usually achieved through ganged switching of the two functions in calibration noise source ignition and indicator scale change.

In 1967, Haroules et al. [2] described a passive circuit which, when introduced at the input of a relative power measuring radiometer, provides an absolute power measurement capability. The input circuit performance is such that the zero position of the output indicator corresponds to zero degrees Kelvin. The effective blackbody temperature of a noise source coupled to the input of the radiometric receiver is read directly in degrees

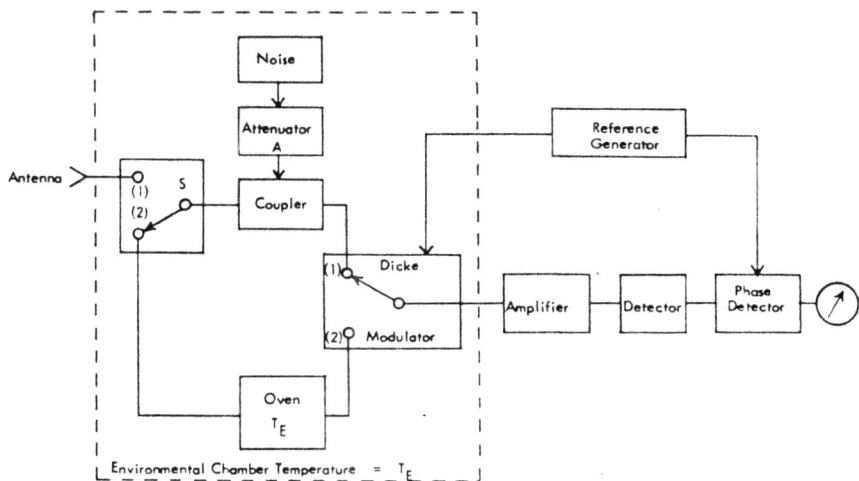

FIG. 6. Functional block diagram of absolute temperature measurement mode.

Kelvin at the output indicator. A simplified functional block diagram showing the interconnection of RF components with the typical Dicke modulator to achieve this absolute radiometric mode is shown in Fig. 6. The concept of operation is predicated on the fact that any noise signal presented at the input port (1) of switch S will be attenuated by the RF losses associated with the passive circuitry in the signal path from the input port of the switch to the input port of the Dicke modulator. In addition, these same passive RF circuit components will radiate noise as a consequence of the fact that they are at a thermometric temperature above absolute zero. This radiated noise will be combined with the attenuated signal and presented to the signal input port (1) of the Dicke modulator. Since the attenuation of the input signal can be characterized as a change in the gain of the total system, and the radiated noise of these components as a noise bias term, the desired calibration of the output indicator directly in degrees Kelvin can be accomplished by a single adjustment of attenuator A. Though the laboratory adjustment of attenuator A requires a known source of noise power (cold load), a standard reference source of this type is not required for field operation of the device.

III. MICROWAVE RADIOMETER APPLICATIONS

The number and diversity of microwave radiometer applications today is truly amazing when one recalls that this instrument technique is just 27 years old in 1970. Exploited first by the young science of radio astronomy, the power of this instrument has demonstrated its ability to explore the unknown and provide many historic discoveries. Only a small fraction of the knowledge gained has been anticipated. It was at first frustrating to learn that the brightness temperature of the sun, measured at low frequencies, was more than $1,000,000°K$ rather than the anticipated $6000°K$, that the radio noise from our own galaxy was markedly different from the anticipated blackbody radiation, that spatially discrete sources of intense radio energy were present in space and could not be identified with optically observed sources. The impact of these and several other discoveries dispelled the frustration and replaced it with a humble admission of the depth of our ignorance. Today, the unanticipated in galactic radio astronomy is considered routine. We now know we are at the dawn of a new era in astronomy, accumulating new knowledge on which we will build a new and deeper understanding in the years ahead.

Spawned by the pioneering and explosive enthusiasm of the young science of radio astronomy, the improvement in microwave radiometric sensor capabilities has been equally startling. The temperature measurement sensitivity achieved today at a wavelength of 3 mm matches the best that could be done 20 years ago at a wavelength of 20 cm. In these two decades, the measurement capability at 20 cm has been improved by more than two orders of magnitude. The instrument of 20 years ago filled an entire room, consuming nearly one kilowatt of power. Its counterpart today is the size of a matchbox and requires less than five watts of input power.

Paralleling the need for improved sensor capability has been the need for larger antennas of improved surface tolerance to provide greater angular resolution at short wavelengths. With the combined improvements in angular resolution and temperature measurement capability, the latest telescopes are now able to probe the familiar neighbors of our own solar system, mapping the surface of the sun and the moon and measuring the thermal radiation characteristics of the planets.

Signals received from space by radio telescopes at wavelengths shorter than 3 cm are contaminated by atmospheric attenuation and noise. This interferring source of noise to the radio astronomers has become the signal for the microwave meteorologist. The water vapor resonance at 22.235 GHz and the complex resonant line structure of molecular atmospheric oxygen near a wavelength of 5 mm became sources of intense investigation beginning in the early 1960s.

The pattern of historical development in the application of the radio telescope was just the reverse of what one might have anticipated. From the early observation of galactic radio noise, followed by the observation of our neighbors in the solar system, the radio telescope in just the past decade has been pointed with greater interest at the planet earth. The first step in this direction has been to obtain an improved understanding of the physical processes in our earth's atmosphere. The microwave radiometer offers the possibility to obtain a global picture of temperature and water vapor distributions. To see, in clear air, large concentrations of water vapor and associated temperature gradients, to detect air mass motion by listening to the radio signal from ozone as it moves like a tracer element through the jet streams, to predict where the clouds will form and whether there will be a major storm, are the exploration frontiers of the microwave meteorologist. A significant amount of fundamental research has already been accomplished. The initial tests of this new potential capability have been scheduled for satellite experiments to be performed in the decade of the 1970s.

In much the same way that the science of radio astronomy spawned the new science of microwave meteorology, the knowledge gained and the associated techniques of microwave meteorology have led to a host of new and exciting areas of application. Our limited objective in the discussion which follows will be to provide a brief summary of present-day microwave radiometer applications in selected areas of research. The exploratory nature of present-day research in each area emphasizes the newness of the measurement instrument. Exploitation of the knowledge gained will be paced by our ability to understand what we learn and our ability to develop useful systems based on that understanding.

A. RADIO ASTRONOMY

By the late 1950s, antenna and receiver technologies had extended radio telescope operation to wavelengths shorter than 3 cm. The millimeter region became the pioneering challenge of the 1960s. Early research in the 3 cm to 3 mm region was concentrated primarily on the determination of the spectral indices of the discrete radio sources which had been previously detected at the longer wavelengths. Cosmic radio maps were extended to the shorter wavelengths and an intensive search for a tenuous isotropic radiation was initiated by several research groups. The existence of this radiation was predicated on the theory of cosmic evolution which postulates an initial primordial explosion [3]. Verification requires an absolute temperature measurement at several wavelengths, one of the most difficult of all radiometric measurements. Partial success has been achieved by the efforts of several investigators, placing the present brightness temperature of this tenuous radiation at a value approximately 3° above absolute zero. Experiments of this type and others in the millimeter wavelength range are complicated by the far fainter cosmic signal levels received from space in comparison with the energy received in the UHF and lower microwave region. Of equal significance is the marked increase in atmospheric attenuation and emission at millimeter wavelengths [4].

As we approached the last year of this decade, one might have said that the radio astronomy discoveries in the millimeter region were insignificant in comparison with the excitement produced at longer wavelengths by the discovery of quasi-stellar objects, pulsars, interstellar OH, helium, and several resonant lines of hydrogen. As is so frequently the case in radio astronomy, the picture suddenly changed when in December 1968, a research group at the University of California [5] announced the detection of radio emission from ammonia molecules in the interstellar medium at a wavelength of 1.25 cm. The line emission was observed at a frequency corresponding to the inversion transitions of the $J = 1$, $K = 1$ rotational levels in the vibrational ground state of the NH_3 molecule. The emission region was of small angular extent, displaced to the south from the direction of the galactic center by approximately 3 arc minutes. The 20-foot diameter millimeter wavelength antenna at the Hat Creek Station of the University of California's Radio Astronomy Laboratory was used with a Dicke type radiometer in which the reference source at the input comparison port of the switch was provided by an off-axis antenna beam, displaced approximately 20 arc minutes from the boresight of the antenna main beam. This technique is used extensively in millimeter wave radio astronomy for the observation of spatially discrete sources of radiation. The continuum radiation of the earth's atmosphere is intercepted by both antenna beams, and hence its contribution results in a null output through the power subtracting action of the Dicke switch at the radiometer input.

As astrophysicists were just beginning to ponder the significance of this startling discovery, the same research group at Berkeley announced the discovery of microwave emission from water vapor in the interstellar medium. This discovery was announced within less than 30 days following publication of their detection of interstellar ammonia. The microwave emission from water vapor was associated with the $6_{16} \rightarrow 5_{23}$ rotational transition. It was observed in several directions in space, one toward Sgr B2, the Orion Nebula, and also in the direction of the source W49. The emission of H_2O in Sgr B2 was in the same direction (celestial coordinates) in which emission from interstellar ammonia had previously been discovered. The H_2O radiation was impressively intense, producing an antenna temperature of $14°K$ when observed from the Orion Nebula and an antenna temperature of at least $55°K$ from the direction of the W49 source. As a consequence of the small angular size of the source regions, the Berkeley group suggested that the brightness temperature of the source in W49 might be as great as $1000°K$. Measurements performed at the Naval Research Laboratory, Maryland Point Observatory, a few weeks following the announcement by the Berkeley group, provided confirmation by measuring an antenna temperature of approximately $1000°K$ from the W49 source, indicating that the actual brightness temperature may be even higher. The antenna at the Naval Research Laboratory, Maryland Point Observatory is 85 feet in diameter. The measured increase in antenna temperature over that obtained by the 20-foot diameter telescope at the University of California Hat Creek Observatory closely followed the ratio of the two antenna aperture areas. Water vapor emission in the direction of W49 is now the most intense emission line detected in the interstellar medium. One can only imagine what future applications will be made of these exo-atmospheric point sources of coherent radiation, in addition to the astrophysical knowledge they will provide.

With the advent of the space age beginning in 1957, the solar system received increased attention as powerful radio telescopes in the 3 cm to 3 mm wavelength region were pointed toward our nearby celestial neighbors. The moon, of course, has been a prime target. Its surface, shape, and general form are wellknown from optical observations. From infrared observations we know it's surface temperature and how it changes so drastically from lunar night to lunar day. With a radio telescope, we look below the surface. The heat wave from the sun propagates below the surface of the moon producing a temperature distribution at the lower levels which is determined by the flux intensity of the heat wave and the thermal properties of the subsurface material. The electromagnetic radiation originates below the surface and propagates upward and out through the surface. The longer the wavelength, the deeper the source of radiation that is observed. The thermal inertia of the subsurface material is the prime parameter of interest. The significant data is the brightness temperature of a selected region observed during one complete lunation. The technique most frequently employed is to obtain a daily map of the brightness distribution over the entire lunar disk and from these maps construct graphical plots of the brightness temperature as a function of lunation phase. The important characteristics are the observed variations in the amplitude of the brightness temperature and the phase lag in the heating and cooling portions of the lunation cycle throughout a lunar month.

As one might anticipate, the amplitude of lunar brightness temperature variations is barely detectable at a wavelength of 3 cm and is undetectable at meter wavelengths. Radiation at these wavelengths originates several meters below the surface of the moon where the heating and cooling of the surface during lunar day and night (approximately 27 days) have a negligible effect. As we approach millimeter wavelengths, however, amplitude variations in the brightness temperature are easily detected. Variations as great at 200°K are typical at a wavelength of 3 mm, and 100°K at 8 mm. The shorter the wavelength, the greater the similarity of the lunation brightness temperature with the surface temperature observed at infrared wavelengths.

Although the physical reasoning associated with this area of lunar research may appear relatively simple, the associated experimental, as well as analytical, problems represent a significant challenge. Among the several factors to be considered are the ability of the antenna to resolve a selected area of the moon, the effects of surface roughness on the insolation phase, and the contaminating effects of the earth's atmosphere on the received signal. At a wavelength of 8 mm, for example, a 30-foot diameter antenna provides a main beam angle response of approximately 4 arc minutes, corresponding to an area approximately 240 miles in diameter on the subterrestrial point of the moon. The main beam angle of the 120-foot diameter Haystack antenna at the M.I.T. Lincoln Laboratory observes a circular area

60 miles in diameter at this wavelength. Since the moon subtends an angle 30 arc minutes in diameter, as observed from the earth, several of the forward sidelobes of either a 30-foot or 120-foot diameter antenna will intercept the lunar surface. Although the sidelobe levels may be low in comparison to the main beam response, the thermal energy radiated by the moon and received by the sidelobe structure is determined by the integral of the brightness temperature of the moon and the gain of the antenna pattern over the solid angle subtended by the moon.

The lunar surface is, perhaps, the most troublesome in the analysis of the observed data since one of the prime objectives is to relate the phase of the observed thermal radiation to the phase of the insolation or heat wave penetrating into the surface. The complexity of the problem can be seen readily by considering the well-known Tycho crater which has a small central prominence projecting from the crater floor. At lunar sunrise, one wall of the crater is exposed to the solar heat wave, while the other remains in a shadow. The central prominence is partially heated by the radiant energy from the wall exposed to the rays of the sun; the other side of the prominence remains cool in its own shadow. As the sun rises, the crater floor and, ultimately, the prominence are exposed to the direct rays of the sun; and finally the wall on the sunrise side of the crater is heated directly by the sun, shortly after midday. As we proceed toward sunset, the reverse situation occurs. It is readily apparent that the central prominence, as well as the floor of the crater, undergo a more complex cycle of heating and cooling than a flat area of exposed lunar surface material. Translating this situation into other areas which are pock-marked with many small crevices, rills, and craters provides an appreciation for the challenge in this area of exploration. Perhaps the greatest challenge is the development of a unified theory capable of explaining what we see in the optical, infrared, and microwave portions of the spectrum, both actively as well as passively, and the relationship of this understanding to what we will soon know about the physical and chemical characteristics of the actual surface material from in situ measurement. From this point of understanding, we may hope to remotely probe the characteristics of other natural satellites in our solar system.

Radio emission from all of the planets, with the exception of Neptune and Pluto, has now been detected. Antenna resolving power is the most significant instrument limitation. Venus, at close approach to the earth, is only 1 arc minute in diameter. Larger collecting apertures, perhaps in the form of multielement interferometers, may be required before we will be able to obtain detailed maps of the thermal radiation characteristics of these nearby celestial neighbors which remain elusive because of their small angular size.

Atmospheric contamination of the signals received from the planets in the millimeter region suggests that sensor systems at these short wavelengths may be more useful if placed in earth orbit or, possibly, on the moon. An alternative, of course, is to send space vehicles to the planets, equipped with sensors covering the entire electromagnetic spectrum. The first historic venture of this type was the Mariner R Fly-by Mission to Venus in 1962. Similar systems are now planned for the "Grand Tour" of the outer planets in 1978. The obvious advantage of planetary space probe is angular resolving power. The gigantic 250-foot diameter radio telescope at Jodrell Bank in England, for example, provides a resolving power on the lunar surface equivalent to an antenna the size of a quarter in a 100 mile lunar orbit.

Although our moon and the planet Venus have been the subjects of intense investigation during the past decade, exploration of solar radio characteristics has, perhaps, received the greatest emphasis. The need to understand and to be able to predict the occurrence of solar activity, in particular the time and intensity of a solar proton flare, has been emphasized by the era of manned space flight. Beyond the protective sheath of the earth's magnetosphere, these corpuscular streams of high energy particles can have damaging effects on life. The tempo of exploration of solar

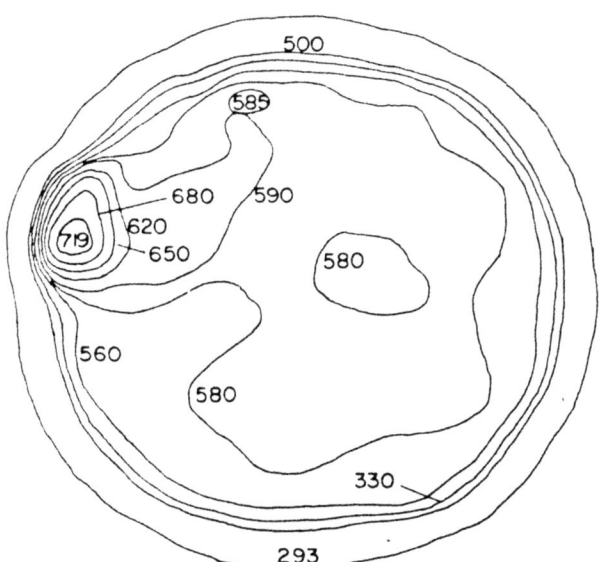

FIG. 7. Sun map at a wavelength of 8 mm taken by the Prospect Hill radio telescope of the Air Force Cambridge Research Laboratories. Temperature contours are in °K/10.

radio characteristics has increased steadily over the past 11-year solar cycle. Today, global communication grid networks transmit data from earth based, optical, and radio telescopes and solar orbiting pioneer satellites to the solar activity forecasting center located in Colorado Springs.

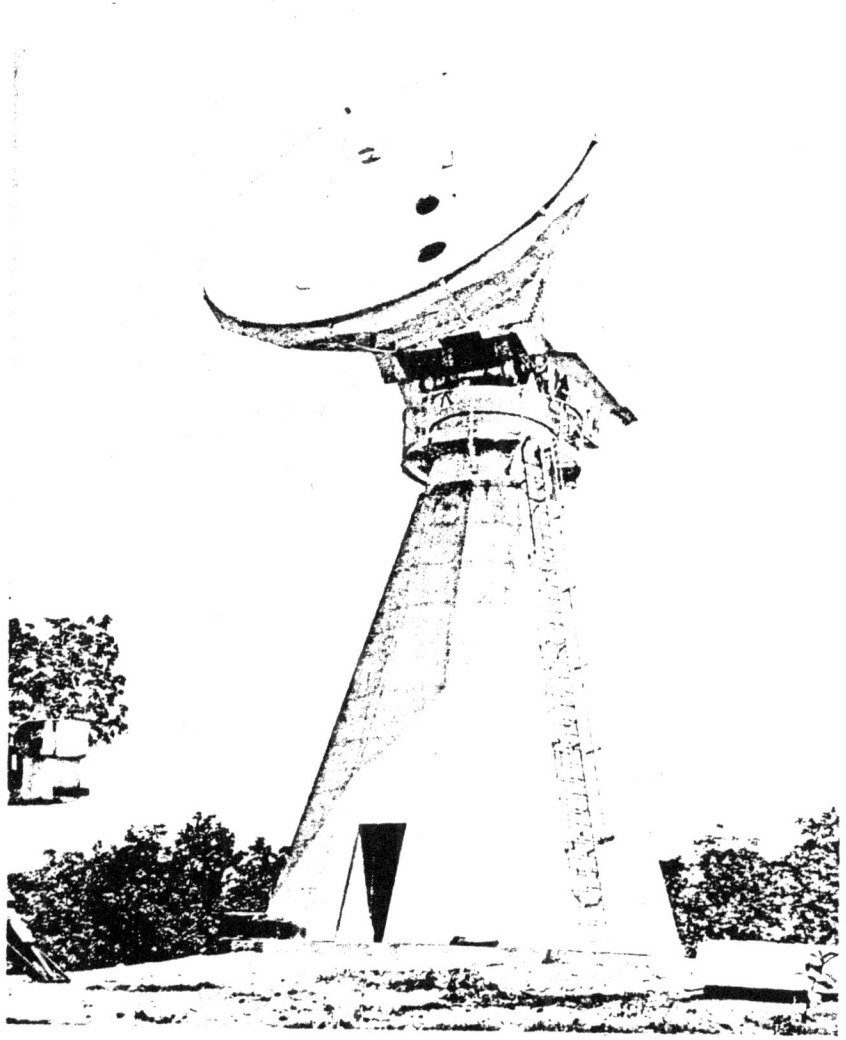

FIG. 8. Air Force Cambridge Research Laboratories, Prospect Hill millimeter radio telescope.

There the data is compiled, reduced, and retransmitted to solar research centers throughout the world. The major effort at several of these research centers is to search for the existence of some characteristic radio signature that can be used as a reliable indicator of a pending major solar event. The search for such a precursor was extended into the millimeter wavelength region in the early 1960s. From prior research, it was known that the high energy proton flares were associated with regions of densely ionized clouds (plage regions) in the vicinity of the solar chromosphere above the optically observed photosphere. It is in this region of the solar atmosphere that the bright flash of the optically observed solar flare occurs at the onset of a major solar event. It was reasoned that at millimeter wavelengths the observed brightness temperature would originate in this same general altitude layer of the solar atmosphere since radio penetration toward the photosphere increases as the wavelength of observation decreases.

Two types of millimeter solar radio telescopes are in use today. One scans a pencil beam across the solar disk in a raster type manner to provide a detailed map of the brightness temperature distribution. The second employs a large antenna beam encompassing the entire solar disk to provide a measure of the integrated brightness temperature as a function of time.

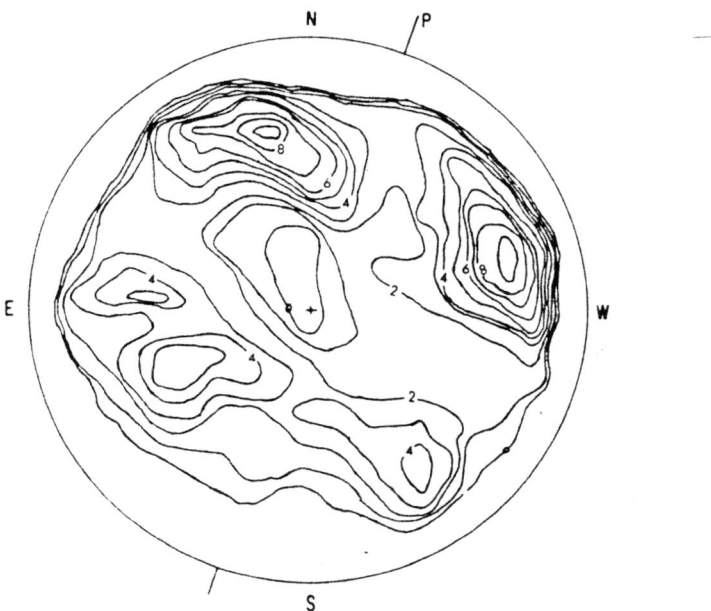

FIG. 9. Aerospace Corporation 3.3 mm sun map for May 25, 1967. The contour plot is in 1% intervals with the reference contour denoted by 0.

The·view of the integrated disk assures that all major flares will be detected and recorded. The mapping instrument runs the risk of missing an event, or at least the onset of an event, unless by sheer coincidence, the antenna beam crosses the area of the flare at the time of occurrence. The mapping

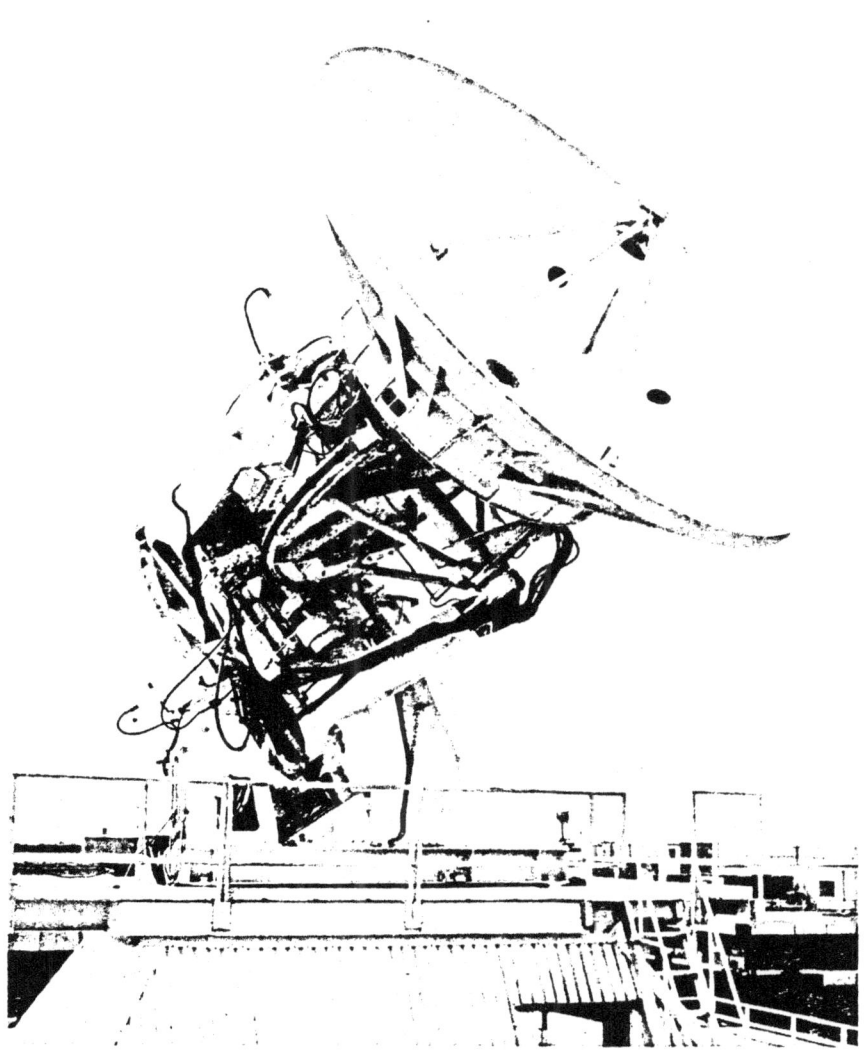

FIG. 10. Aerospace Corporation millimeter radio telescope.

instrument is, of course, far more sophisticated than the instruments which provide an integrated view of the total solar flux intensity. The antenna motion required to achieve the raster scan is accomplished by computer programming of the drive system. The output data is computer-reduced to a map display of brightness temperature contours distributed over the surface of the solar disk. Comparison of these brightness temperature maps, in time sequence, clearly indicates the growth and decay of active regions just above the photosphere. An 8 mm brightness temperature map of this type, obtained by P. Kalaghan of the Air Force Cambridge Research Laboratories (AFCRL) Prospect Hill Observatory, is shown in Fig. 7. A photograph of the (AFCRL) solar radio telescope is shown in Fig. 8. The antenna system is an elevation over azimuth mounted 30-foot diameter cassegrain. Solar brightness temperature maps obtained at a wavelength of 3 mm are shown for comparison in Fig. 9. This 3 mm map was recorded by the Aerospace Corporation radio telescope located in El Segundo, California. A photograph of this equatorially mounted 15-foot diameter cassegrain antenna is shown in Fig. 10.

B. MICROWAVE METEOROLOGY

Microwave meteorology is in part an outgrowth of radio astronomy and more recently has become closely allied with atmospheric propagation studies needed to determine system requirements for earth-space communication links at millimeter wavelengths. One area of millimeter communication systems research is the measurement of atmospheric attenuation and noise characteristics in the "windows" between the water and oxygen resonances. These resonant lines occur at nominal frequencies of 22 and 60 GHz respectively.

A second area of research in microwave meteorology is the measurement of the characteristics of atmospheric resonant lines themselves to capitalize on their unique properties which can be exploited for the purposes of determining the vertical temperature profile, water vapor distribution, and ozone distribution in the earth's atmosphere.

Present-day investigations have produced an interesting cross-breeding between research groups in communication, radio astronomy, and microwave meteorology. Celestial radio sources, for example, offer excellent exo-atmospheric tragets of opportunity for the communicator to investigate atmospheric propagation characteristics by satisfying the requirement for a transmitter of known flux intensity and position located beyond the earth's atmosphere. The validity of such measurements is, of course, predicated on available knowledge concerning the characteristics of these celestial transmitting sources. As a consequence, a substantial portion of our present-day knowledge concerning the millimeter characteristics of celestial radio sources is being provided by research scientists primarily concerned with the communication aspects of the atmospheric medium.

The common denominator in each discipline is the microwave radiometric sensor. The atmospheric medium is, of course, another common denominator. However, depending on the research discipline, the electromagnetic characteristics of the atmosphere may be considered as either signal or noise.

In the discussion which follows, we will consider first the measurement of atmospheric attenuation and noise in the "windows." We will then review present-day exploration of atmospheric gas resonant line structure which is located between the "windows" and forms their boundaries. Two areas of application in which present knowledge concerning atmospheric resonant line structure is being exploited will be reviewed in the third section.

1. *Measurement of Atmospheric Attenuation and Noise*

At wavelengths shorter than 3 cm, electromagnetic waves can be severely attenuated by rain. A possible solution for earth-to-satellite communication systems would be the use of a diversity ground station at each earth terminal. Under conditions of severe attenuation caused by excessive rain at a ground terminal, the communication link could be switched to the diversity ground station. The minimum separation between the two earth based terminals requires a quantitative determination of attenuation statistics. The accumulation and analysis of the required statistical data is currently being pursued at the Air Force Cambridge Research Laboratory, the Bell Telephone Laboratory, and the NASA Electronics Research Center. In lieu of the availability of a man-made satellite to provide a calibrated exo-atmospheric transmitting source, the sun is used as a target of opportunity since it is an exo-atmospheric source of known flux intensity. Because of the relatively low signal level received from the sun even when unattenuated, a radio telescope is required for these measurements. The noise level received from the sun as a function of weather conditions determines the attenuation caused by the intervening atmospheric gases and associated condensation and precipitation products. During nighttime when the sun is not available for absorption type measurements, the radiometric sensor is pointed at elevation angles of interest, and atmospheric effects are derived from observed variations in the sky noise temperature.

When the antenna is pointed at the sun, the antenna temperature T_s is given by

$$T_s = \int \frac{G(\theta, \phi)}{4\pi} \left[T_{s\bar{s}} \exp(-\tau) + (1 - \exp(-\tau))T_{sky} \right] d\Omega \qquad (16)$$

where $G(\theta, \phi)$ is the gain pattern of the antenna in spherical coordinates, τ is the atmospheric opacity, $T_{s\bar{s}}$ is the exo-atmospheric brightness temperature of the sun at the frequency of observation, and T_{sky} is the mean thermometric temperature of the atmosphere along the path of observation.

When the antenna beam is pointed away from the sun toward some cold point in the sky, the observed antenna temperature T_A is given by

$$T_A = \int \frac{G(\theta, \phi)}{4\pi} (1 - \exp(-\tau))T_{sky} \, d\Omega \qquad (17)$$

Now if the observations of T_s and T_A are made in sequence at the same elevation angle by pointing the antenna beam first on then off the sun, the antenna temperature difference between the two measurements will be

$$T_s - T_A = \exp(-\tau) \int \frac{G(\theta, \phi)}{4\pi} T_{s\bar{s}} \, d\Omega_s \qquad (18)$$

where $d\Omega_s$ is the solid angle of the sun.

Introducing the reasonable assumption that the atmospheric absorption coefficient will be essentially constant over the small angle subtended by the sun, the observed antenna temperatures when the antenna beam is pointed toward the sun and then toward the sky can be written in the form

$$T_s = T_{sb} \exp(-\tau) + [1 - \exp(-\tau)]T_{sky} \qquad (19)$$

$$T_A = [1 - \exp(-\tau)]T_{sky} \qquad (20)$$

For clear weather conditions, one can obtain a measure of the atmospheric gas attenuation by assuming a uniform horizontally stratified model atmosphere. Under these conditions, the atmospheric opacity τ can be reexpressed in terms of the opacity observed in the zenith direction τ_0, where

$$\tau = \tau_0 \sec \theta_z \qquad (21)$$

and θ_z is the angle of observation referenced to the zenith direction.

Expressing τ as a function of τ_0 in Eqs. (19) and (20), we can rearrange terms and obtain the expression

$$\frac{T_s - T_{sky}}{T_{sb} - T_{sky}} = \exp(-\tau_0 \sec \theta_z) \qquad (22)$$

from which we obtain the value of the vertical opacity τ_0 in the form

$$T_0 = \left[1\eta \frac{T_{sb} - T_{sky}}{T_s - T_{sky}} \right] \cos \theta \qquad (23)$$

Alternatively, the value of τ_0 can be determined from the slope of the plot of $T_S - T_{sky}$ versus $\sec \theta_z$.

For nighttime observations, a similar method can be applied to determine the atmospheric opacity from the measured sky antenna temperature. The value of τ_0 is obtained from the derivative of Eq. (20) in the form

$$\tau_0 = \frac{1}{T_A - T_{sky}} \frac{d(T_A - T_{sky})}{d \sec \theta_z} \qquad (24)$$

For large values of τ, the assumed value of the mean thermometric temperature of the atmosphere T_{sky} becomes critically important. One can make the general observation that large values of attenuation are more precisely measured in absorption by using the sun as an exo-atmospheric source, while small changes in the value of atmospheric attenuation are more effectively sensed by sky temperature measurements obtained under conditions of relatively low attenuation values.

The determination of atmospheric attenuation characteristics from sky temperature measurements has been extensively used and developed to a high degree of sophistication by the research group at the Air Force Cambridge Research Laboratory, under the direction of E. Altshuler. In addition to equipment located at the Prospect Hill Observatory in Waltham, Massachusetts, the AFCRL group operates a dual frequency measurement

FIG. 11. Air Force Cambridge Research Laboratories dual frequency (15 GHz and 35 GHz) radiometric measurement system located at Mount Hilo in the Hawaiian Islands.

instrument at Hilo in the Hawaiian islands. A photograph of this instrument, which is operated under the direction of K. Wulfsberg of AFCRL, is shown in Fig. 11. The entire equipment enclosure rotates in azimuth, and the cornucopia antenna attached to the side of the enclosure is adjusted in elevation from the operator control console located within the enclosure.

A microwave radiometric sensor system assembled by the Bell Telephone Laboratories for the accumulation of atmospheric attenuation statistics at wavelengths of 8 mm and 2 cm is shown in Fig. 12. The

FIG. 12. The Bell Laboratories sun tracker in Holmdel, New Jersey is used to tune in on sun signals at two radio frequencies. A 5 × 9-foot metal mirror automatically follows the sun in its daily path across the sky. Other electronic equipment processes the signals and records the results. The apparatus is gathering data on the effect of rain on the signals received. (2/68.)

5 × 9-foot plane reflector is attached to an equatorial mount. The declination angle of the reflecting plane is adjusted so that the sun's rays are reflected into the 4-foot aperture conical horn reflector antenna. The reflector is driven in the hour angle coordinate by a clock mechanism which assures that the sun's rays are continuously reflected into the aperture of the conical horn reflector. Throughout the observing period, the antenna beam is scanned on and off the sun at a 1 Hz rate, with an angular excursion of 2.6°, by mechanically tilting the reflecting plane in the decli-

nation angle coordinate. Automatic operation is another unique feature of the instrument. It is preprogrammed several days in advance and provides continuous accumulation of data with unattended operation. Nighttime observations of sky noise are included in the observing program sequence.

The Propagation Studies Branch at the NASA Electronic Research Center, under the direction of L. Roberts, undertook a similar series of measurements, beginning in 1967. The ERC atmospheric propagation measurement system is shown in Fig. 13. Simultaneous observations are obtained at wavelengths of 3 cm, 2 cm, 1.5 cm, and 8 mm. The microwave radiometric sensors at each wavelength are installed on individual equatorially mounted 5-foot diameter searchlights. Several modes of oper-

FIG. 13. NASA Electronics Research Center atmospheric research instrumentation. These sun tracking instruments are used to obtain atmospheric attenuation data at four wavelengths between 3 cm and 8 mm. The antenna systems are equatorially mounted 5-foot searchlights.

ation are included in these radiometric systems: antenna beam switching, absolute temperature, and exploded antenna beam comparison. The exploded beam comparison mode provides a differential temperature measurement between two on-axis antenna beams, one narrow beam boresighted on the sun and a much larger beam which obtains a negligible contribution from the sun.

Although the prime objective of measurements of this type is to determine the statistics of atmospheric attenuation and noise needed for the design of future earth-space communication links, the knowledge gained is useful in understanding the physical processes of the atmosphere. Simultaneous observations at several wavelengths selected to exploit the wavelength dependent atmospheric absorption coefficient offers the potential for remotely mapping clouds and weather fronts. Penetration to the rain cores within clouds is accomplished at the longer wavelengths. Detection of high-altitude variations in water vapor, prior to condensation, can be sensed at the shorter wavelengths. The instrumentation required for these measurements is markedly similar to that described above for communication systems research. From the routine daily accumulation of attenuation statistics may evolve similar instruments used by microwave meteorologists to determine the "why" of these statistics.

2. Absorption and Radiation by Atmospheric Gases

The atmospheric gases which provide a significant interaction with microwaves are water vapor, oxygen, and ozone. Water vapor has strong absorption lines at 1.35 cm and 1.63 mm, as well as several strong lines at submillimeter wavelengths. Van Vleck [6] calculated the magnitude of the 1.35 cm line and the contribution from all other lines. Comparison of his results with laboratory measurements by Becker and Autler [7] and Ho et al. [8] and with atmospheric observations by Straiton and Tolbert [9] showed a substantial discrepancy. By adjusting the contribution from all other lines, the water vapor absorption formulas summarized by Barrett and Chung [10] represent the best available approximation for the 1.35 cm line at temperature near 300°K.

Oxygen has a complex spectrum, consisting of a band of resonant lines in the 5 mm wavelength range and an isolated line at 2.5 mm. Line frequencies and bandwidths have been measured in the laboratory at pressures up to 1 atmosphere by Artman and Gordon [11] and Anderson et al. [12]. Direct measurement of atmospheric absorption by oxygen has been made by several investigators. The measurement results and computations were reviewed by Meeks and Lilley [13] and, recently, by Westwater and Strand [14].

The interaction of ozone with microwaves is weak in comparison with either oxygen or water vapor. Several resonant lines, however, are present throughout the entire 3 cm to 3 mm wavelength region. Gora [15] has calculated the frequencies and intensities for all significant lines of the rotational spectrum of ozone at frequencies below 2700 GHz.

The application of microwave radiometric sensing of the water vapor resonant line is presently being exploited because it offers the unique ability to yield a measurement of tropospheric water vapor in the presence of clouds. Although a satellite borne water vapor sensor would be effective only over oceans, oceans cover more than half the earth's surface and are the spawning grounds of major storms. Microwave radiometric sensing of the water vapor resonance under clear weather conditions permits the measurement of integrated water vapor abundances and spatial size distributions.

Interest in the oxygen resonant line characteristics near a wavelength of 5 mm has been stimulated by the fact that microwave radiometric sensing may provide the only remote sensing technique capable of measuring atmospheric temperature profiles in the presence of clouds. This would be of considerable importance to global data collection for numerical weather prediction. This technique offers the potential capability of measuring the temperature profile from the lower troposphere well into the mesosphere.

Microwave radiometric measurement of the atmospheric ozone distribution has progressed at a slower pace than studies of either oxygen or water vapor. Instrument technology, rather than meteorological interest, has set the pace in this area of research. Ozone plays an important role in the organic and inorganic chemistry of the surface of the earth. Through a filtering action, it absorbs a lethal part of the ultraviolet radiation from the sun, thereby making life possible on the surface of the earth. Ozone is also an important factor in our climatology, establishing the balance between exo-atmospheric radiation incident on the earth and the outgoing radiation from the earth, as a consequence of its particular absorption characteristics in the ultraviolet and infrared regions of the spectrum. Knowledge of the atmospheric ozone distribution in the altitude range from 15 to 60 km, obtained on a global scale, offers the possibility of measuring air mass circulation as a consequence of the fact that ozone in the lower portions of the atmosphere may be considered an inert gas and its global distribution with time is, in large, determined by the horizontal motion and interaction of major air masses.

As a preface to a review of the current status of exploration of atmospheric gas resonant characteristics, it will be helpful to recall the relationship between antenna temperature and the effective brightness temperature of an observed source of radiation. Since the atmosphere throughout most of the 3 cm to 3 mm wavelength region is semitransparent, the equation of radiative transfer can be used to relate the brightness temperature $T_B(\nu)$ at the frequency ν to the atmospheric composition and the temperature $T(s)$ along the line of sight and to the brightness temperature T_E of the background medium beyond the atmosphere. The equation of radiative transfer is expressed in the form

$$T_B(\nu) = T_E \exp\left[-\tau(\nu)\right] + \int_0^{s_{max}} T(s) \exp\left[-\int_0^s \alpha(\nu, s)d(s)\right] \alpha(\nu, s)d(s) \quad (25)$$

In Eq. (25) $\tau(\nu)$ is the total opacity of the atmosphere and $\alpha(\nu, s)$ is the absorption coefficient. Inspection of Eq. (25) shows that the observed brightness temperature in any given direction is the sum of the background radiation and the radiation emitted at each point along the path of observation, each component attenuated by the intervening atmosphere. The antenna temperature observed by a microwave radiometric sensor looking into the atmosphere is, therefore, primarily determined by the atmospheric absorption coefficient and temperature along the path of observation. Since the integral of the product of the exponent and the absorption coefficient in Eq. (25) determines the contribution of the thermometric temperature along the path of observation, to the observed antenna temperature, it has become customary to refer to the value of this integral as a "weighting function."

  a. Oxygen. The microwave spectrum of the oxygen molecule results from fine structure transitions in which the magnetic moment assumes various directions with respect to the rotational angular momentum of the molecule. The unpaired spins of two electrons produce the magnetic dipole moment of oxygen. Van Vleck [16] was the first to develop the expression for the frequency, pressure, and temperature dependence of the oxygen absorption coefficient. This early work was reviewed by Meeks and Lilley [13] in 1963, and later by Gautier and Robert [17] in 1964, and Lenoir [18] in 1968.

  The complex of oxygen lines, in particular the atmospheric absorption coefficient as a function of frequency and altitude of observation, is shown in Fig. 14 (Meeks and Lilley). The general form of the weighting functions for selected frequencies, as computed by Meeks and Lilley, is shown in Fig. 15.

  It is important to note that the expression for the oxygen absorption coefficient has been derived from quantum mechanical considerations in which the value of certain constants has been empirically selected to provide the best agreement with experimental data. If it were possible to directly measure the absorption coefficient for all possible meteorological conditions of interest, the quantum mechanical approach could be dis-

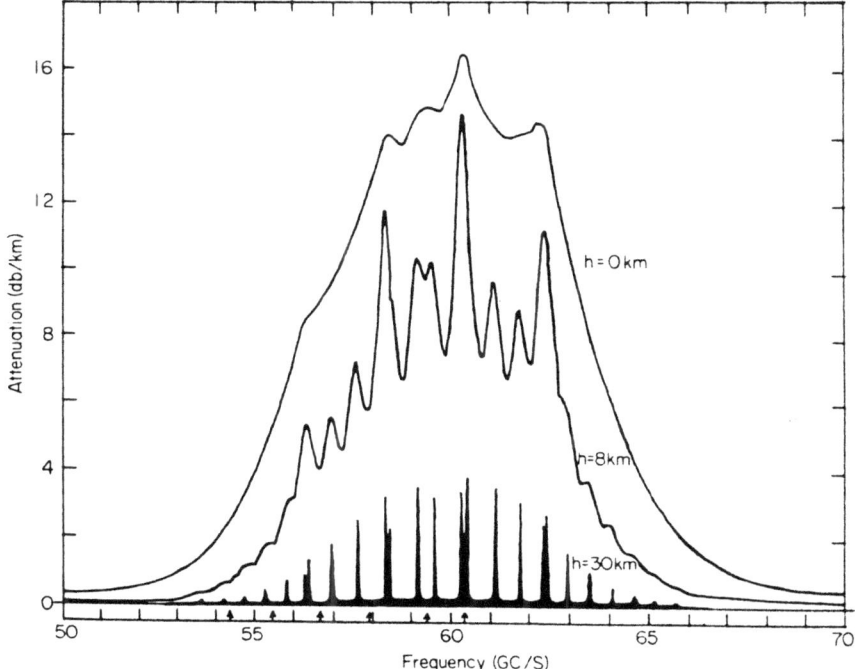

FIG. 14. The computed attenuation coefficient $\gamma(\nu)$ for air at three representative heights. This figure shows that the individual oxygen lines completely overlap at sea level, partly overlap at 8 km, and are resolved at 30 km. (Meeks and Lilly [13].)

pensed with. Much of the present-day research has been concentrated on the direct measurement of oxygen line profile characteristics to provide an improvement in knowledge concerning the absorption coefficient values. Laboratory measurements have been performed by Stafford and Tolbert [19] at the University of Texas, and balloon measurements by Lenoir [18] at the Massachusetts Institute of Technology.

Reber *et al.* [20] of the Aerospace Corporation recently reported a very detailed analytical study of this problem, supported by extensive measurements performed in a high-altitude aircraft. Their published values are, perhaps, the most comprehensive and complete available today. Measurements, utilizing the sun as a source, were made at six discrete altitudes ranging from sea level to 13.7 km. These measurements covered the frequency range from 52 to 68 GHz (see Fig. 16). The more than 1500 independent attenuation measurements were used to calculate new

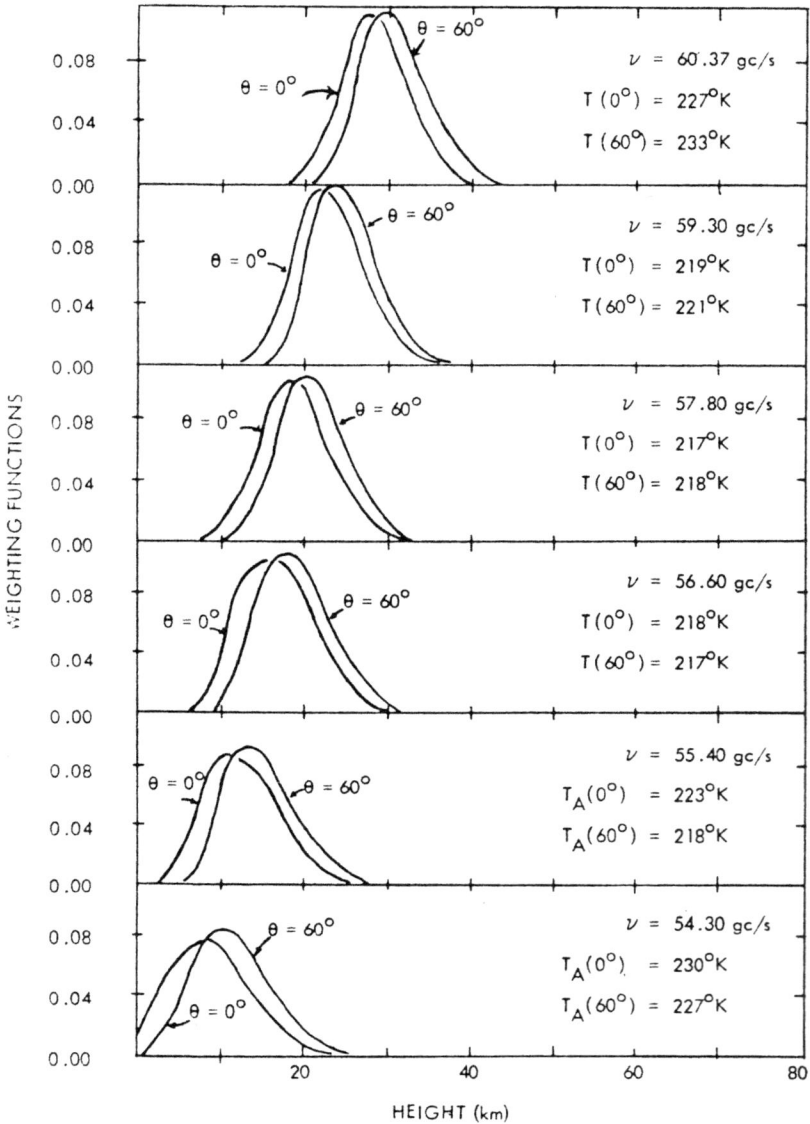

FIG. 15. Weighting functions for determination of the brightness temperature as a weighted average of the kinetic temperature distribution. Weighting functions are shown for six representative frequencies. (Meeks and Lilley [13].

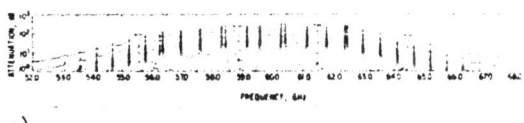

a)

b)

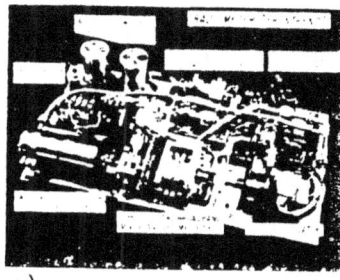

c)

d)

FIG. 16. Key features of the investigation of molecular atmospheric oxygen characteristics performed by the Aerospace Corporation. (a) The computed resonant profile characteristics of atmospheric oxygen as a function of the observers altitude over the frequency range from 52 to 68 GHz. (b) Five-millimeter wavelength radiometric sensor antenna system assembled for the aircraft measurement program. (c) RF portions of the 5 mm radiometric receiver used in the aircraft measurement program. This unit was physically located directly behind the antenna. (d) The 5 mm radiometric sensor installed in the aircraft used in the measurement program. The antenna can be seen behind the quartz window installed in the skin of the aircraft, just forward of the wing.

values for the Van Vleck line-broadening coefficients. Zenith attenuations were computed utilizing these new coefficients over the frequency range 48 to 72 GHz and for several altitudes from 0 to 25 km. In addition, both horizontal attenuation rates and tangential attenuations through the atmosphere were computed for several altitudes.

   b. Water Vapor. There is only one water vapor resonance in the 3 cm to 3 mm wavelength region. This occurs at a wavelength of 1.35 cm. A second line occurs at a slightly shorter wavelength of 1.6 mm. There are a great number of strong lines at wavelengths shorter than 1 mm. The atmospheric opacity expressions for the 1.35 cm line were first developed by Van Vleck [6]. These were further refined by Barrett and Chung [10]. They obtained relatively good agreement between theory and experiment by combining the Van Vleck and Weisskopf [21] line shape with a nonresonant term which corresponds to contributions from the far wings of all of the lines at other frequencies. The most recent analytical and experimental work has been by Staelin [22] and Gaut [23]. Their experimental measurements were performed using a 28-foot aperture antenna at Lincoln Laboratory in Lexington, Massachusetts with a 5-channel microwave radiometer. The five selected frequencies were observed simultaneously to obtain an absorption profile using the sun as the background source. The accuracy of these profile measurements was of the order of 0.02 dB, or less than 5% of the total opacity.

   The measurement of the water vapor concentration in the atmosphere by microwave radiometric techniques is complicated by the fact that the water vapor resonance at 1.35 cm is semitransparent. It is also relatively broad since most of the atmospheric water vapor is located at altitudes below 10 km, where pressure broadening effects dominate the line shape. Consequently, vertical sounding of the atmospheric water vapor distribution from satellite orbit will be contaminated by radiation emitted from the earth's terrain or ocean surface, as well as by clouds. For these reasons, the exploration of the potential of this particular remote sensing capability is being pursued in several areas. These include a more precise determination of the characteristics of the absorption coefficient, the effect of clouds on received signal characteristics, and the radio emission characteristics of the ocean's surface. Observations obtained over the oceans will be the most useful since the emissivity of the ocean is approximately one-half that of the land, thereby providing an adequate differential temperature contrast. Observations over land areas will provide little, if any, distinguishable signal since the temperature of the lower atmosphere, where water vapor is most adundant, is close to the earth ambient at the surface.

   c. Ozone. The principal quantity which determines the transmission coefficient of the atmosphere and the emission or effective brightness temperature of the atmosphere due to ozone is the absorption coefficient as a function of frequency and altitude. Those parameters which determine the variation of the absorption coefficient with altitude are temperature, pressure, and ozone concentration. Gora [15] calculated the frequencies and intensities for all significant lines of the rotational spectrum of ozone at frequencies below 2700 GHz. The values of molecular constants were determined from the frequencies of ozone lines in the microwave spectrum, measured by Trambarulo et al. [24] and by Hughes [25]. The average half-width of the ozone lines in the 9.6-μ band, as determined by Walshaw [26], was used by Gora to calculate what he termed the maximum absorption coefficient. He also used the Lorentz line shape factor in this calculation since this function is a valid approximation at the low pressures typical of those regions of the atmosphere where ozone is concentrated.

Atmospheric measurement of the 36 GHz line in absorption by Mouw and Silver [27], the 37.8 GHz line absorption and the 30.1 GHz line in emission by Caton et al. [28], and 23.8 GHz line in emission by Barrett et al. [29] provided direct verification of the existence of these lines and their relative intensities as predicted by Gora. The experimental measurement of a far more intense line at 101.7 GHz reported by Caton et al. [30] offered the first opportunity to directly measure line profile characteristics to an accuracy sufficient for inversion. The measured line width of this 3 mm transition is in excellent agreement with the predicted width as inferred from the infrared measurements by Walshaw and the laboratory measurement of the 37.8 GHz line by Caton et al. [28].

It will be helpful at this point to briefly review the methods used in the analysis of the observed ozone data to determine the percentage of ozone concentration relative to the total air content at various altitudes [31]. Certain disciplinary relationships are significant since there is a common use of terminology in the concept of "weighting functions." The approach to ozone data analysis provides a simplification in measurement instrument requirements since an absolute temperature measurement is not required at a single frequency in order to deduce the ozone concentration. The difference temperature between measurements made at two frequencies is used to infer the concentration in an atmospheric layer. A precise measure of the temperature difference between the two frequencies is required; however, the absolute value of either is not required.

The relationship between the ozone absorption coefficient α and significant variable parameters may be expressed in the form

$$\alpha(\nu) \propto T^{-5/2} \frac{N_{03}}{\Delta\nu} \left[\frac{(\Delta\nu)^2}{(\nu - \nu_0)^2 + (\Delta\nu)^2} \right] \qquad (26)$$

where N_{03} is the ozone concentration, $\Delta\nu$ is the line half-width, ν is the frequency of observation, and ν_0 the line frequency. The line width is proportional to pressure, which is in turn proportional to the product of the total number of air molecules N_T and a temperature term $T[\exp(-5/2 + \beta)]$, where β is in the range 0.5 to 1.0. The absorption coefficient can, therefore, be rewritten in the form

$$\alpha(\nu) = DT\frac{N_{03}}{N_T}\left[\frac{(\Delta\nu)^2}{(\nu - \nu_0)^2 + (\Delta\nu)^2}\right]. \qquad (27)$$

The constant of proportionality D includes molecular constants and geometrical factors associated with the path of observation through the atmosphere. The term in brackets is defined as a "single frequency weighting function" W_ν. A "difference weighting function" $W_{\nu_1} - W_{\nu_2}$ is defined in the form

$$W_{\nu_1-\nu_2} = B\left[\frac{(\Delta\nu)^2}{(\nu_1 - \nu_0)^2 + (\Delta\nu)^2} - \frac{(\Delta\nu)^2}{(\nu_2 - \nu_0)^2 + (\Delta\nu)^2}\right] \qquad (28)$$

where β is a constant which normalizes the difference frequency weighting function to unity. Inspection of Eq. (28) indicates that the maximum value of $W_{\nu_1-\nu_2}$ occurs when

$$(\nu_1 - \nu_0)(\nu_2 - \nu_0) = (\Delta\nu)^2 \qquad (29)$$

Since the observed brightness temperature at any single frequency of observation is proportional to the integration of the absorption coefficient along the path through the atmosphere, the ozone concentration can be derived from a difference temperature measurement at two frequencies which define the altitude limits of the observed layer. A graphical plot of six single frequency weighting functions is shown in Fig. 17a. The corresponding difference weighting functions are shown in Fig. 17b. The three difference frequency weighting functions shown in solid line are obtained from the differences of the paired sets of the four single frequency weighting functions, also shown in solid line in Fig. 16a. The single frequency weighting functions at 2 and 20 MHz (shown dotted) which combine to form the difference frequency weighting function at 6.3 MHz (dotted) show the degree to which difference frequency weighting functions at 2, 6.3, and 20 MHz tend to overlap and thereby provide interdependent samples of the atmosphere.

The development of the weighting functions in Fig. 17 demonstrates the dual use of data obtained at intermediate observing frequencies; i.e., observational data obtained at 6.3 MHz can be used for the upper and middle difference frequency weighting functions, as well as for both the middle and lower difference frequency weighting functions.

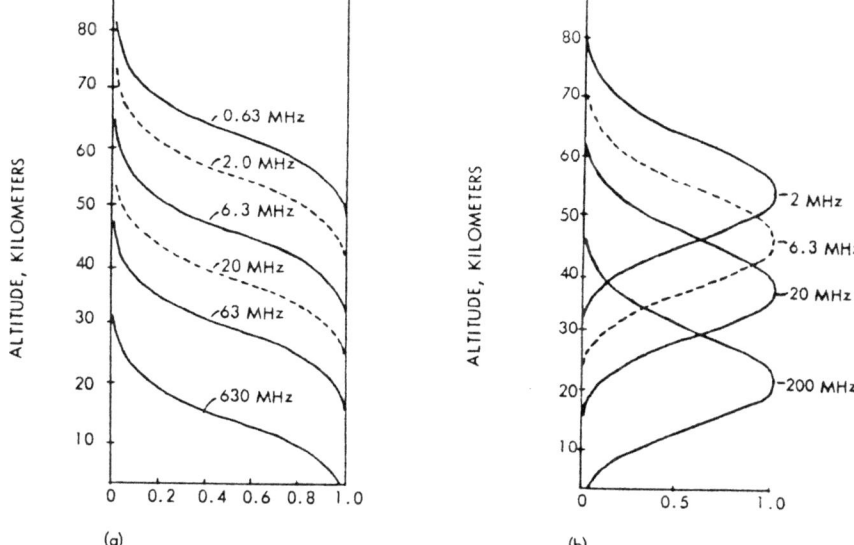

FIG. 17. Weighting functions of atmospheric ozone. (a) Normalized single frequency weighting functions W_ν versus frequency displacement from the
line frequency. (b) Normalized difference frequency weighting functions $W_{\nu_1-\nu_2}$ versus frequency and altitude.

The natural limit of the difference weighting function half-width is
not immediately apparent from the example of atmospheric ozone weighting functions shown in Fig. 17. This natural limit is shown graphically
in Fig. 18. Recalling that the observed temperature at any frequency of
observation is proportional to the integral of the absorption coefficient
over the ray path of observation, a factor $C(h)$, representing all terms in
the integrand other than those in the weighting function, takes the typical
form shown in Fig. 18a for standard ARDC model atmospheric temperature and pressures and for values of atmospheric ozone concentration as
calculated by Hunt [32]. The form of $C(h)$ is primarily determined by the
factors outside the bracket in Eq. (26). From Eq. (28), it is apparent that
the width of the weighting function $W_{\nu_1-\nu_2}$ is determined by the ratio
$(\nu_1 - \nu_0)/(\nu_2 - \nu_0)$. The width of $W_{\nu_1-\nu_2}$ is constant for values of the ratio.
A plot of this ratio versus the width of the difference frequency weighting
function decreases to a minimum of about 14.5 km as the difference
$(\nu_2 - \nu_1)$ decreases to zero (see Fig. 18b). This indicates that as the two frequencies of observation required to define a difference frequency weighting
function approach the corresponding "frequency of the weighting func-

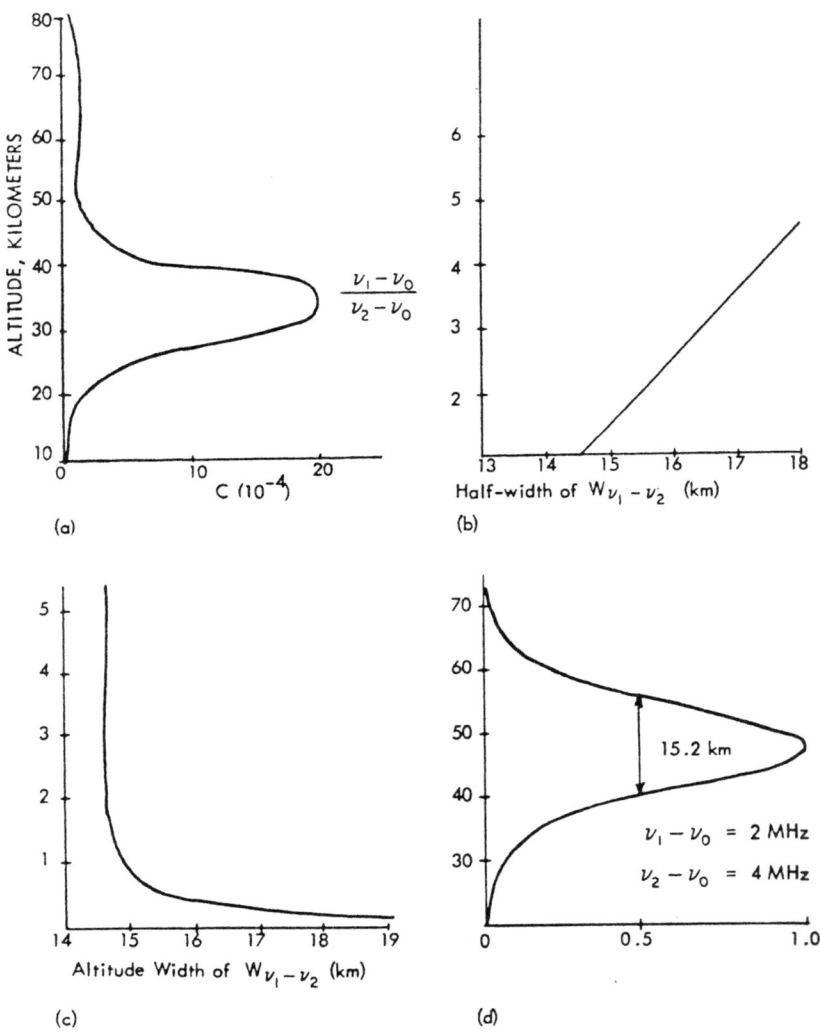

FIG. 18. Half-width limits of ozone difference frequency weighting functions. (a) $C(h)$ vs height (h) in km for standard ARDC model atmosphere and O_3 concentration. (After Hunt [32].) (b) Width of difference frequency ozone weighting function vs $(\nu_1 - \nu_0)/(\nu_2 - \nu_0)$. (c) Normalizing factor B vs width of the difference frequency weighting function $W_{\nu_1 - \nu_2}$. (d) Normalized weighting function $W_{\nu_1 - \nu_2}$ vs altitude (where $2\nu_0 =$ width determined by the atmospheric pressure at the altitude of the peak response).

tion," the minimum half-width of the weighting function becomes 14.5 km. The minimum width of the difference weighting function for useful data can also be seen directly if one plots the value of the factor β in Eq. (28) as a function of the width of the difference frequency weighting function, since this normalizing factor is directly indicative of the energy contained under the area of the weighting function when plotted as a function of altitude and amplitude. The graph of β versus the width of $W_{\nu_1 - \nu_2}$ is shown in Fig. 18c. It is evident from this graph that the minimum usable width of a difference frequency weighting function is approximately 15 km. A graph of a difference frequency weighting function near this optimum width of 15 km is shown in Fig. 18d. The corresponding frequencies of observation for the single-frequency weighting functions are displaced from the line frequency by 2 and 4 MHz respectively. Comparison of the shape and width of this weighting function (Fig. 18d) with the weighting function shown graphically by the dotted line in Fig. 17b indicates that a weighting function developed from observing frequencies displaced 2 MHz and 20 MHz from the center frequency of the line provides nearly the same definition of the atmospheric layer as that provided by the combination of frequencies displaced 2 MHz and 4 MHz from the line center. This tends to suggest the value of using intermediate frequencies of observation to perform a dual role in the derivation of ozone concentration for adjacent weighting functions.

It is apparent from the foregoing that, at most, four independent samples of atmospheric ozone concentration can be obtained. The minimum half-width of each sample layer will be of the order of 15 km. A measure of the ozone concentration in any selected layer is computed from measured temperature differences at two frequencies.

The microwave radiometer used by Caton et al. [30] was a double conversion superheterodyne. First intermediate frequency amplification was provided by three travelling wave tubes in cascade with an instantaneous bandwidth of 2 GHz centered at 3 GHz. The input signal to the second converter was coupled from an interstage transmission line between the second and third traveling wave tubes. Six second intermediate frequency amplifiers were provided in the form of five contiguous filters, each 10 MHz wide, and one filter covering the entire 50 MHz band. A seventh broadband 2 GHz response was derived from the output of the third traveling wave tube. Absorption measurements were performed using the sun as a background source. The comparison load in the Dicke mode of operation was provided by a gas discharge noise source, fed through a servo controlled attenuator to the comparison port of the ferrite modulator. The control signal for the servo loop was derived from the broadband (2 GHz) channel. The servo control loop performed the function of stabi-

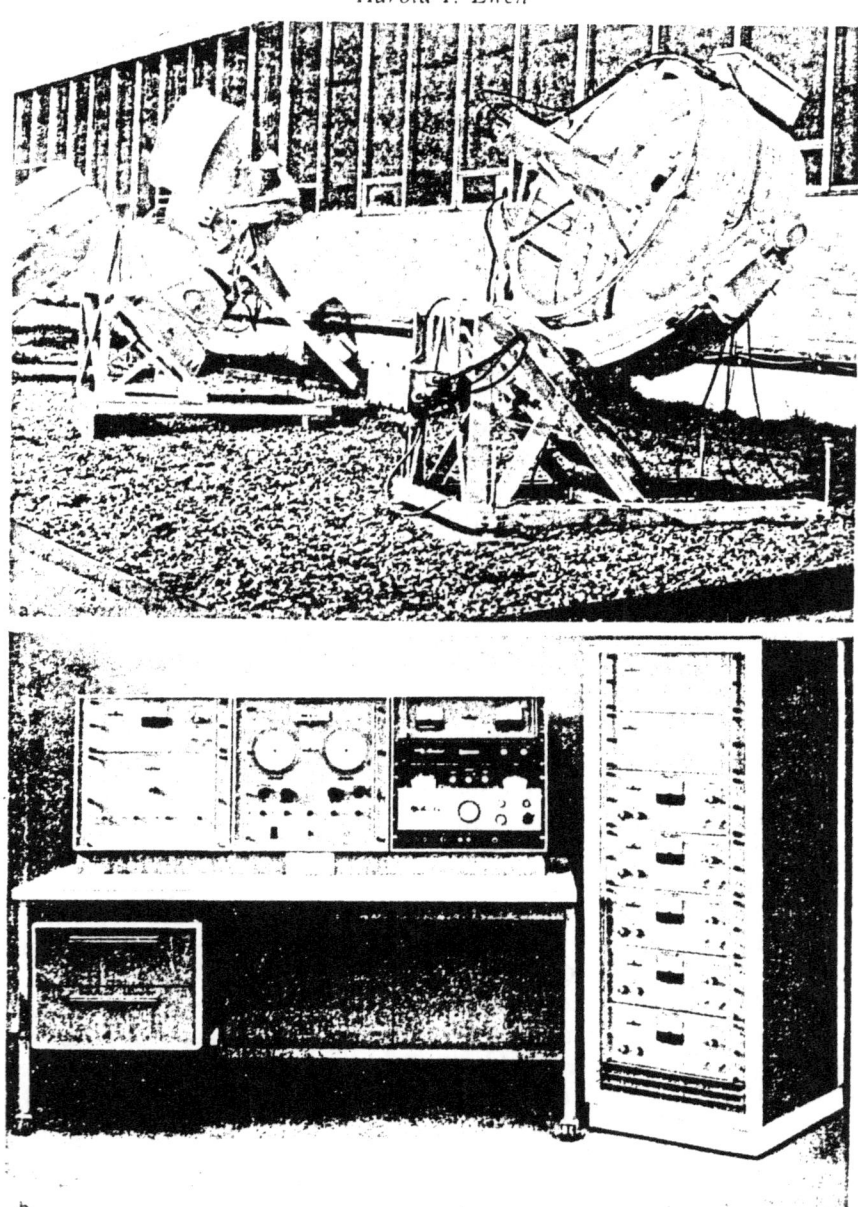

FIG. 19. NASA Electronics Research Center atmospheric ozone radiometric
sensor used in the initial detection of the resonant line at 101.7 GHz.
(a) Equatorially mounted five-foot diameter searchlight antenna. (b)
Antenna and radiometric signal processing control console.

lizing the output from the various channels by discriminating against small variations in the observed sun antenna temperature produced by clouds drifting through the antenna beam during the period of observation. Even under clear weather conditions, variations in the sun antenna temperature as great as $100°K$ were frequently observed at this wavelength. These broadband variations appeared in all channels when the noise feedback loop was inoperative. The antenna was an equatorially mounted 5-foot searchlight. Tracking of the sun was provided by a synchronous clock mechanism. Photographs of the antenna and control console are shown in Figs. 19a and b respectively.

3. *Related Areas of Application*

The absorption characteristics of molecular atmospheric oxygen offer great opportunity for exploitation. The attenuation experienced from sea level along a vertical path through the atmosphere is nearly 300 dB at resonant line frequencies near 60 GHz. Satellite-to-satellite communication at these frequencies would be free of man-made noise originating at the earth's surface. A communication link of this type would also be undetectable at the earth's surface. Communication links between high-altitude aircraft, operating at frequencies in the wells between resonant lines, would enjoy the same benefits.

The ability to view the earth from satellite orbit with a radiometric sensor and see a uniformly bright mantle in this wavelength region has suggested the possibility of an earth vertical sensor more accurate than an IR horizon scanner.

Another possible application is the ability to remotely sense regions of clear air turbulence (CAT) in the forward flight path of supersonic high altitude aircraft. A millimeter wave radiometer, tuned to the oxygen wavelength band, may provide this capability. Temperature anomalies appear to be associated with CAT regions. The range at which a "millimeter wave thermometer" is projected forward of the aircraft along its flight path can be adjusted by selecting the wavelength of observation, capitalizing again on the wavelength dependence of the oxygen absorption coefficient.

The satellite earth-vertical sensor and the CAT detector concept are discussed in greater detail in the sections which follow. These applications are typical examples of the exploitation of knowledge currently being gained and applied to improve present capabilities through new concepts and techniques.

a. An Earth Vertical Sensor. The most common method for passive remote sensing of the earth vertical from satellite orbit is predicated on the symmetry and stability of the earth's infrared horizon about the local satellite vertical. The performance of an IR horizon scanner is determined by natural limitations. It has been suggested [33] that the molecular atmospheric oxygen mantle might offer a superior reference for earth vertical sensing. Although the system concept of rim cutting used in IR horizon definition could be applied to sensing of the molecular atmospheric oxygen horizon mantle, the antenna aperture size to obtain an equivalent pencil

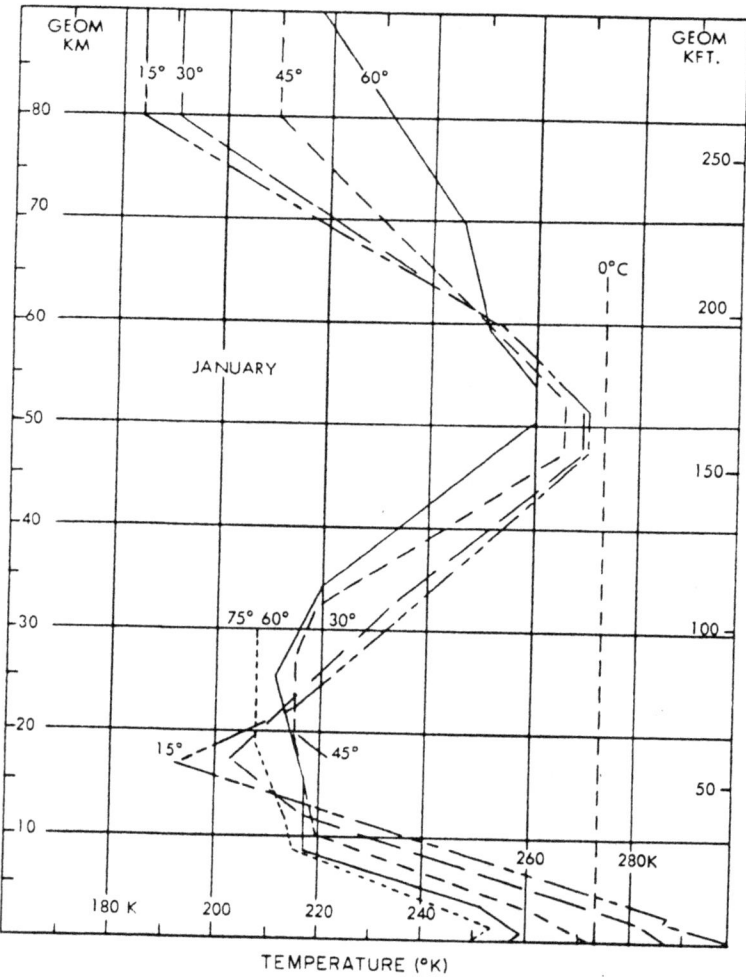

FIG. 20. Typical atmospheric temperature vs height profiles observed in January at latitudes of 15, 30, 40, 60 and 75° north.

beam would be unreasonably large for satellite application. Fortuitously, the thermal radiation characteristics of molecular atmospheric oxygen negate the need for a rim cutting technique.

The relationship between the oxygen emission spectrum and the temperature as a function of altitude above sea level shows that the observed emission is frequency selective and represents the average temperature in an atmospheric layer of air approximately 7 to 10 km deep. The mean

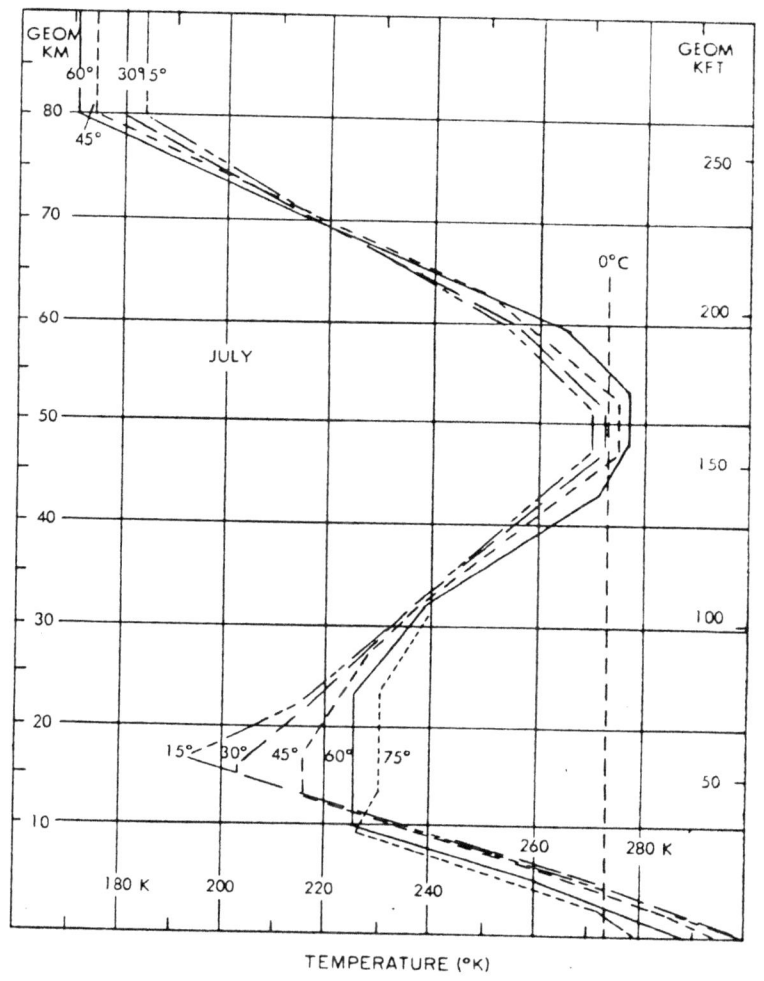

FIG. 21. Typical atmospheric temperature vs height profiles observed in July at latitudes of 15, 30, 45, 60 and 75° north.

height of the observed layer is determined by the frequency of observation and the observation angle relative to the nadir. The basic concept of a molecular atmospheric oxygen vertical sensor is predicated on selection of an observing frequency which provides thermal sensing of the atmospheric temperature at an altitude (determined also by the observing angle relative to the nadir) at which a near uniform global temperature distribution is anticipated.

Typical temperature height profiles for the month of January as a function of latitude are shown in Fig. 20. Corresponding temperature height profiles for the month of July are shown in Fig. 21. With the exception of the reported high temperatures during winter in northern latitudes, one would anticipate temperature variations of approximately $\pm 10^\circ$K from the poles to the equator in the altitude range from 25 to 35 km, independent of season. Referring to the temperature-height profiles (Figs. 20 and 21), it is of interest to note that the temperature difference between widely separated latitudes is less at higher altitudes. It should be noted that atmospheric temperature data above 25 km is quite sparse; and in particular, above 35 km (balloon altitude) is obtained by isolated rocket probes. The general form, however, of the temperature-height profiles suggests the efficacy of observing a near uniform global temperature mantle at altitudes above 25 km.

As previously described, the absorption characteristics of molecular atmospheric oxygen are such that one can select a frequency of observation to obtain a temperature sounding at any desired altitude from sea level to approximately 75 km. The "weighting function" for a particular frequency of observation describes the altitude interval-temperature contribution for that frequency. As shown in Fig. 15, the depth of the weighting functions tends to increase with altitude. All weighting functions shown in Fig. 15 are associated with frequencies located between oxygen line resonant frequencies; i.e., they are located in line "wells" as opposed to line "cores." It is of interest to note, in reference to Fig. 15, that the mean altitude of the weighting function increases with the angle of observation off the nadir direction. The increase for a 60 arc degree zenith angle is of the order of 5 km as shown in Fig. 15. The mean altitude of the weighting function increases sharply as one approaches a zenith angle of the horizon as viewed from orbit. At 60.8 GHz, for example, the altitude of the weighting function increases 11 km over the altitude that would be probed at the same frequency in the nadir direction. This feature is of considerable advantage in system design since the selection of an operating frequency such as 60.8 GHz in the "well" between line "cores" eases restraints on frequency stability, while at the same time providing a weighting function altitude in the 25 to 30 km range, where temperature variations with latitude are of the order of $\pm 10°K$.

The radiometric mode of operation for a molecular atmospheric oxygen earth vertical sensor is frequently referred to as thermal centroid sensing. This passive microwave technique was developed shortly after World War II and has undergone several generic advancements since that time. A major area of application has been in the design of radiometric sextants used for the navigation of mobile vehicles (primarily ships and submarines) under foul weather conditions. Celestial radio sources such as the sun, moon, and radio stars are used in these applications. In the radiometric sextant mode, the projection of the antenna beam on the celestial sphere is invariably much larger than the solid angle subtended by the celestial source. The position of the source in orthogonal coordinates about the radio boresight of the antenna is determined by comparing the power received by dual antenna beams in either coordinate. The two beams are usually displaced to provide a 3 dB response on the antenna boresight axis. The general form of the angle-tracking error function at the output of the radiometric receiver in either coordinate is independent of the method of angle sensing; however, the achievable signal-to-noise ratio is critically dependent on the mode of operation and, hence, determines the rms angle tracking accuracy.

The development of the expression for the angle-tracking accuracy achieved in a radiometric sextant configuration will be helpful in the analysis of the earth vertical sensor concept, since the latter is a degenerate case of radiometric sextant thermal centroid tracking. In the earth vertical sensing application the antenna beam angle is smaller than the solid angle of the celestial source (earth).

The form of the angle error function in either coordinate in the radiometric sextant mode is shown in Fig. 22. Note that the solid angle of the source is smaller than either antenna beam projection on the celestial sphere. The two antenna beams are identical, each providing a half-power response on the boresight axis. A characteristic "S-curve" is developed as the source passes through the line of centers of the two beams and the boresight axis. The S-curve is derived by subtracting the power received by antenna beam B from that received by antenna beam A. Noise fluctuations associated with the thermal noise characteristics of the target and inherent receiver noise are superimposed on the S-curve. The rms angle tracking accuracy can be derived from the following geometrical considerations.

The slope of the straight line connecting the peaks of the S-curve with the target source on boresight is

$$\text{Straight line slope} = \frac{(P_K \text{ to } P_K)S}{\theta_A} \qquad (30)$$

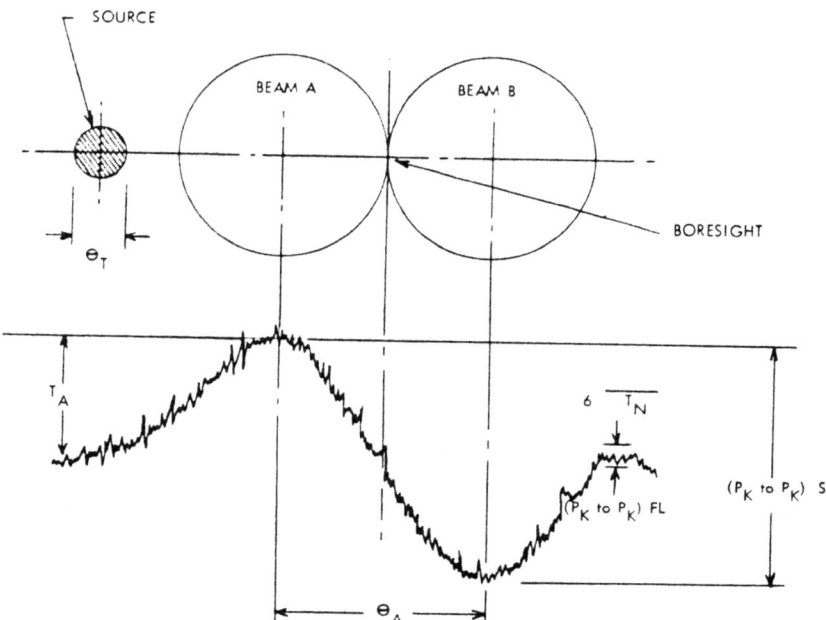

F_{IG}. 22. General form of a radiometric sextant angle error curve (S-curve) derived by two antenna beams whose power patterns intersect at their half-power points on 'the boresight axis of the antenna system. The angle error curve is obtained by subtracting the power received by antenna beam *B* from that received by antenna beam *A* as a celestial source (sun, moon, or radio star) passes along the line of centers of the two antenna beams. The "figure-of-merit" of the curve is defined as the ratio of the peak-to-peak amplitude $(P_K$ to $P_K)$ S to the peak-to-peak level of receiver noise fluctuations $(P_K$ to $P_K)$ FL, superimposed on the curve.

where $(P_K$ to $P_K)S$ is the peak-to-peak amplitude of the S-curve and θ_A is the angular separation of the two peaks of the S-curve. For a $\cos^2$ antenna aperture illumination, the slope is $\pi/2$ times greater than the straight line slope, or:

$$\text{S-curve slope} = \pi/2 \frac{(P_K \text{ to } P_K)S}{\theta_A} \tag{31}$$

The rms angle tracking error at boresight $\overline{\delta\theta_A}$, multiplied by the slope of the S-curve at boresight, is equal to the rms value of the fluctuating noise component superimposed on the S-curve, or

$$\pi/2 \frac{(P_K - P_k)S}{\theta_A} \overline{\delta\theta}_A = \frac{(P_k \text{ to } P_K)FL}{6} \tag{32}$$

where $(P_K \text{ to } P_k)FL$ is the peak-to-peak noise fluctuation level superimposed on the S-curve. Therefore

$$\overline{\delta\theta}_A = \frac{\theta_A}{3\pi M_S} \tag{33}$$

where

$$M_S \equiv \frac{(P_k \text{ to } P_k)S}{(P_k \text{ to } P_k)FL} \tag{34}$$

M_S is defined as the figure-of-merit of the S-curve and represents a measure of the signal-to-noise ratio associated with the angle-tracking error function. The angle sensing mode of operation must be selected to maximize M_S.

The equation for the figure-of-merit M_S may be re-expressed in radiometric system parameters by recalling that the peak-to-peak amplitude of the S-curve is twice the observed antenna temperature when the source is centered in either antenna beam and that the peak-to-peak noise fluctuation level is 6 times the rms radiometric sensitivity (twice 3σ), i.e.,

$$(P_K \text{ to } P_K)S \approx 2T_A \approx T_B\left(\frac{\theta_T}{\theta_A}\right)^2 \rho \tag{35}$$

$$(P_K \text{ to } P_K)FL \approx 6\overline{\Delta T_N} \approx \frac{\sqrt{\pi}(E-1)T_0}{\sqrt{\hat{\beta}t}} \tag{36}$$

where

T_B = the brightness temperature of the source

θ_T = the angle subtended by the source

θ_A = the antenna beam angle (3 dB points)

$\overline{\Delta T_N}$ = rms sensitivity of the radiometer, °K

ρ = the antenna aperture efficiency

F = the receiver noise figure

β = the receiver predetection bandwidth

t = the receiver postdetection integration time constant

$T_0 = 290°K$

The foregoing analysis can be extended to the earth vertical thermal centroid sensor mode by noting that for this case the antenna beam is smaller than the target, as shown in Fig. 23. If the squint angle between

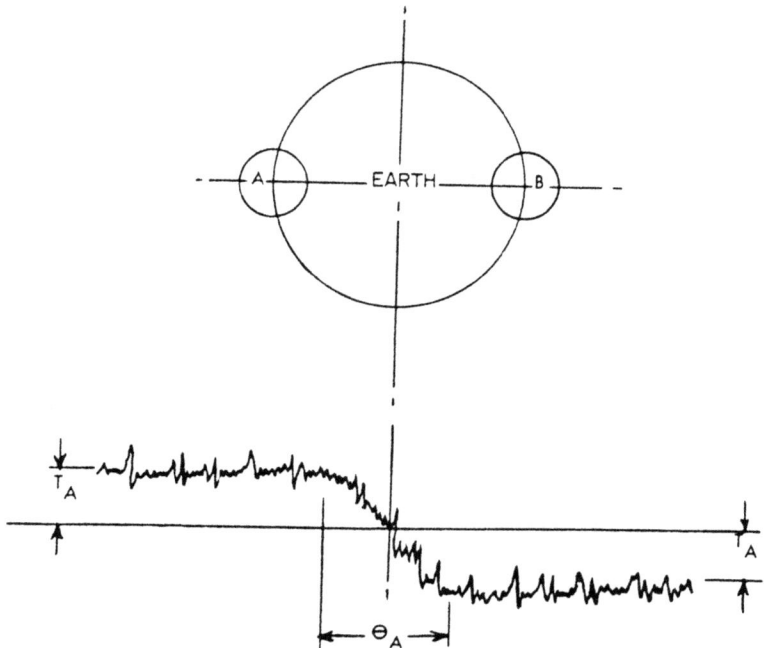

FIG. 23. General form of the angle error curve developed by a satellite earth vertical sensor when operating in the thermal centroid sensing mode. The earth vertical direction is coincident with the null crossover point of the angle error curve. This is a degenerate case of the radiometric sextant mode shown in Fig. 22.

antenna beam A and B is adjusted to provide simultaneous interception of opposing earth horizon lines on the corresponding antenna boresight axis, then an S-curve will be developed as the two antenna beams are scanned across the earth, as shown Fig. 23. In this case, the antenna temperature T_A will be the brightness temperature of the earth oxygen mantle at the frequency of observation (for an antenna aperture efficiency of 100%).

Inspection of this special case leads to the conclusion that Eq. (33) is directly applicable to the determination of the rms angle tracking accuracy associated with the crossover or null point of the S-curve. As a typical example of anticipated performance, we may assume that the mantle temperature of oxygen at an altitude of 30 km will be in the order of 230°K. Hence, for a 3°K rms sensitivity achieved with a postdetection time constant of 0.1 sec, the figure-of-merit of the S-curve will be

$$M_S = \frac{2T_A}{6\,\overline{\Delta T_N}} = \frac{230}{9} = 25.5 \tag{37}$$

For a 10° antenna beam angle, the anticipated rms angle tracking accuracy is therefore 2.5 arc minutes (rms).

A photograph of the first radiometric sensor assembled for satellite flight to evaluate the efficacy of this concept is shown in Fig. 24. This

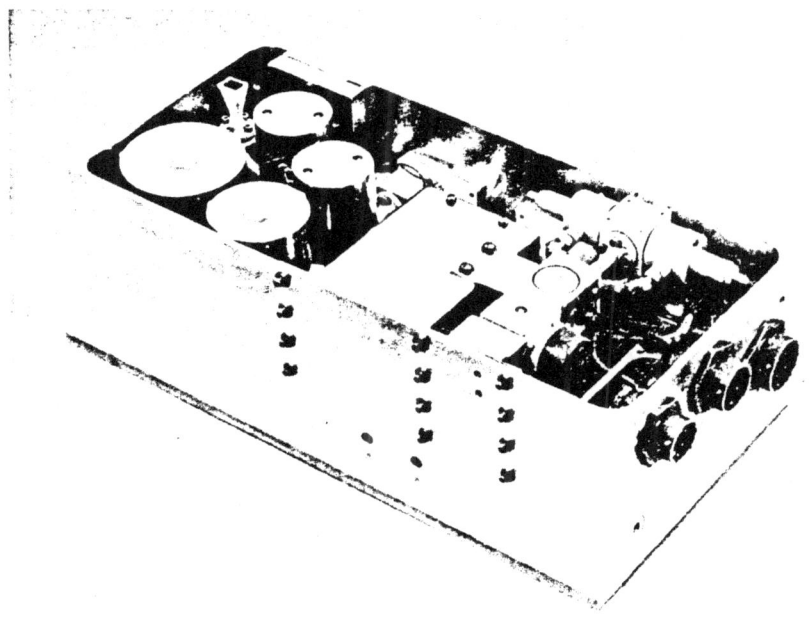

FIG. 24. A 5-mm radiometric sensor developed by the Air Force Cambridge Research Laboratories. This unit was launched into earth orbit in July 1967.

unit was developed by the Air Force Cambridge Research Laboratories under the direction of J. Aarons and D. Guidice. A modified absolute temperature mode of operation was used, with the output indicator zero adjusted to correspond to an input signal temperature of 250°K. The objective of the experiment was to determine if the atmospheric temperature observed from satellite orbit at a frequency of 60.8 GHz was consistent with the predicted atmospheric model. The observed temperature as a function latitude was a prime measurement objective. The radiometer was launched into a near-circular polar orbit on July 27, 1967. Prior to stabilization, the satellite tumble rate of 1 rpm was easily detected as the single antenna beam scanned across the earth and sky. Difficulty was encountered in achieving vehicle stabilization with the desired attitude. The satellite ultimately stabilized in an upside-down attitude, pointing the sensor antenna into space. For a period of three months, the unit faithfully recorded that the observed temperature was near zero. This was an important pioneering experiment, however, since it demonstrated that radiometer instrument technology at 5 mm was ready for the challenge of exploration from space at this wavelength.

b.   Detection of Clear Air Turbulence. The ability to detect temperature anomalies in the forward flight path of a high-altitude aircraft at a wavelength of 5 mm is predicated on the large dynamic range of atmospheric absorption coefficients which are available over a relatively small wavelength range near 5 mm as a consequence of the resonant line profile characteristics of molecular atmospheric oxygen. The intensity and range to an atmospheric temperature anomaly along the forward flight path of an aircraft is sensed by operating at two or more frequencies selected to provide atmospheric absorption coefficients at the flight altitude in the range from 0.1 to 1.0 dB/km.

Assuming an ideal pencil beam antenna pattern (the antenna pattern unity over an angle corresponding to 3 dB antenna beamwidth and zero elsewhere), the expression for the observed antenna temperature [see Eq. (25)] takes the form

$$T_A(\nu) = \int_0^\infty T(S) \left[\exp - \int_0^{S_{max}} \alpha(\nu, S)\, dS \right] \alpha(\nu, S)\, dS \qquad (38)$$

where

$\alpha(\nu, S)$ = volume absorption coefficient at frequency ν and range s.

$T(S)$ = thermometric temperature of the atmosphere at range s.

The integrand is the product of two factors. The first is the temperature distribution along the ray path, the second is a space-dependent function of the attenuation coefficient along the ray path. The second factor is largest for those regions nearest the antenna and exponentially decrease as the distance from the element of atmosphere located in range interval ds becomes progressively farther from the antenna. Thus, this factor emphasizes spatial elements of the temperature distribution at ranges near the antenna and provides a decreasing contribution to antenna temperature for those elements well removed from the antenna. Because of this spatial selection property, the second factor in the integrand is frequently referred to as the "horizontal weighting function" of the temperature distribution along the ray path.

It is apparent from Eq. (38) that the contribution to the observed antenna temperature from any region along the ray path is determined by the value of the weighting function which is in turn determined by the value of the absorption coefficient at the frequency of observation. The temperature contribution from a region well forward of the antenna can be made to contribute a significant portion to the overall antenna temperature by observing at a frequency with a relatively large value for the absorption coefficient. Selection of the observing frequency is based on knowledge of the frequency dependence of α_{ν} at the flight altitude.

Probing atmospheric temperatures along the forward flight path of an aircraft can be accomplished by a multichannel (multifrequency) radiometer in which the channel frequencies are selected to provide the desired combination of α_{ν} values required for detection of temperature anomalies ahead of the aircraft. A minimum of two frequencies of observation are required.

To illustrate the range capability of the temperature sensing system of this type, we will rewrite Eq. (38) for the case in which the absorption coefficient is constant along the ray path. This is a reasonable assumption for a horizontal path. The antenna temperature for this condition is

$$T_A(\nu) = \alpha_{\nu} \int_0^{\infty} [\exp - \alpha(\nu, S)] T_s \, ds \qquad (39)$$

We now note that the exponential factor is the only range dependent term in the weighting function. Its maximum value occurs when the range $s = 0$. For all other values of s, the weighting function steadily decreases, thereby reducing contributions from regions at large range values. If we define the distance at which the exponential factor is 1% of its maximum value as S_M. i.e.,

$$S_M = \frac{1}{\alpha_{\nu}} \ln 100 \qquad (40)$$

then regions at distance S greater than S_M contribute less than 1% to the total antenna temperature, and S_M defines the range interval which contributes 99% to the observed antenna temperature. Typical values of S_M and corresponding values of α (in dB/km) are: $\alpha = 0.1$, $S_M = 200$ km; $\alpha = 0.5$, $S_M = 40$ km; and $\alpha = 2.0$, $S_M = 10$ km.

To obtain a quantitative picture of anticipated performance, let us consider a 5 mm radiometric system mounted in an aircraft with its antenna beam pointing along the horizontal flight path. If the ambient temperature along the flight path is T_f and a temperature anomaly described by a function $\Delta T(S)$ is present in the forward range interval S_1 to S_2, the antenna temperature sensed by the radiometer will be

$$T_A(\nu) = T_f + \int_{S_1}^{S_2} \alpha(\nu)\, \Delta T(S)[\exp - \alpha(\nu)S]\, dS \tag{41}$$

Notice that in the absence of a temperature anomaly $\Delta T(S)$, the antenna temperature is simply the ambient temperature at the flight altitude T_f. When a temperature anomaly is included, the limits of the second integral are only over the region where the anomaly is present since the integrand

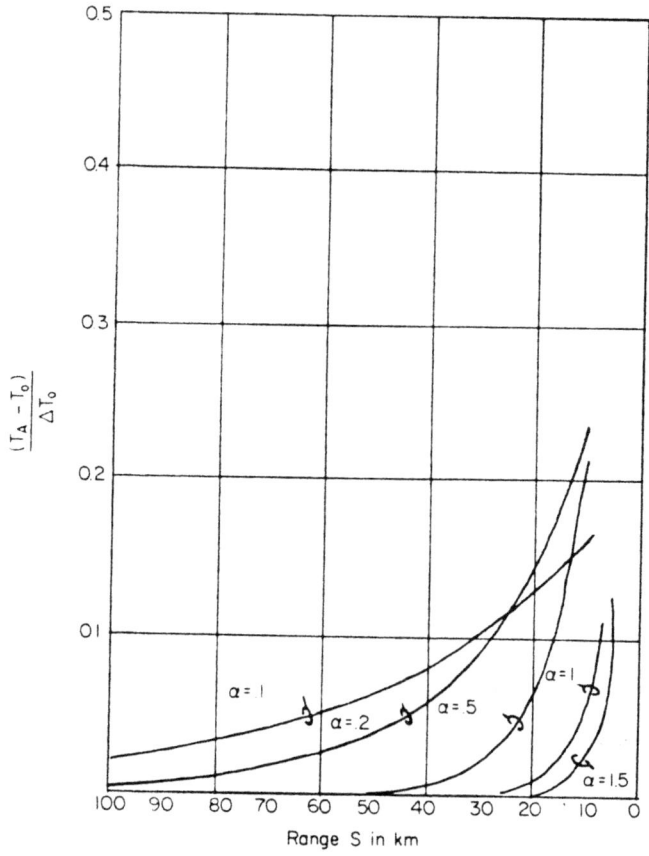

FIG. 25. The normalized multifrequency response of a radiometric sensor to a step temperature anomaly of ΔT_0, 10 km in extent, at a horizontal range S from the sensor. The desired atmospheric absorption coefficient α in the range from 0.1 to 1.5 dB/km determines the corresponding frequencies of observation. For a practical system, the observing frequencies can be confined to a relatively narrow bandwidth by operating in the vicinity of a resonant line of an atmospheric gas, such as molecular oxygen.

is zero elsewhere. As an example, let $\Delta T(S)$ equal a constant value ΔT_0 over the range interval from S_1 to S_2, and zero elsewhere. In this case the antenna temperature is

$$T_A(\nu) = T_I + \exp\left(-\alpha_\nu S_1\right)[1 - \exp\left(-\alpha_\nu (S_2 - S_1)\right]\Delta T_S \qquad (42)$$

Thus, the presence of the temperature anomaly appears as a change in antenna temperature about the ambient T_I.

As an illustrative example of the magnitude of the anticipated change in antenna temperature as a function of range and observing frequency (α value), consider the case of a $10\,°C$ temperature anomaly, 10 km in extent, in the interval range S_1 to S_2. Figure 25 shows a graphical plot of the change in antenna temperature relative to the ambient temperature at the flight altitude, as a function of the range to the temperature anomaly, for selected values of α from 0.1 to 1.5 dB/km. If we assume that the sensor has a temperature sensing capability of $0.5\,°K$ (indicative of present capability), it is apparent from Fig. 25 that the temperature anomaly would be observed at a distance of 59 km at an observing frequency for which $\alpha = 0.1$ dB/km, 42 km for $\alpha = 0.2$ dB/km, and 22 km for $\alpha = 0.5$ dB/km.

Referring to Fig. 25, it is of interest to note the response characteristics of the various channels to the assumed temperature anomaly. Although initially insensitive to the disturbance, the $\alpha = 0.2$ dB/km channel subsequently responds very quickly and, at a range of 25 km, provides an output signal which exceeds the signal level in the $\alpha = 0.1$ dB/km channel. A similar "crossover point" for the channel pair $\alpha = 0.5$ dB/km and $\alpha = 0.1$ dB/km occurs at a range of 8 km. The presence of these "crossover points" between individual channels offers an additional range indicator for the anomalous temperature region.

To demonstrate that the results are not critically dependent upon the temperature profile of the assumed discontinuity, a similar analysis can be applied to two other forms of temperature anomaly. In one case, we will assume a linear transition from temperature T_I to $T_I + 10\,°C$ and in the other an exponential transition from T_I to $T_I + 10\,°C$. For each case, we will assume a half amplitude width of 10 km, corresponding to the first case considered. The anticipated antenna temperature change versus range to the anomalous temperature region resulting from these distributions are shown in Figs. 26 and 27, respectively, for comparison with the case described by Fig. 25. It is of interest to note that overall signature characteristics are the same for all three cases. This indicates that the basic range sensing capability is not critically dependent on the temperature profile of the anomaly but rather on the fundamental radiative properties of molecular atmospheric oxygen as a function of frequency and altitude

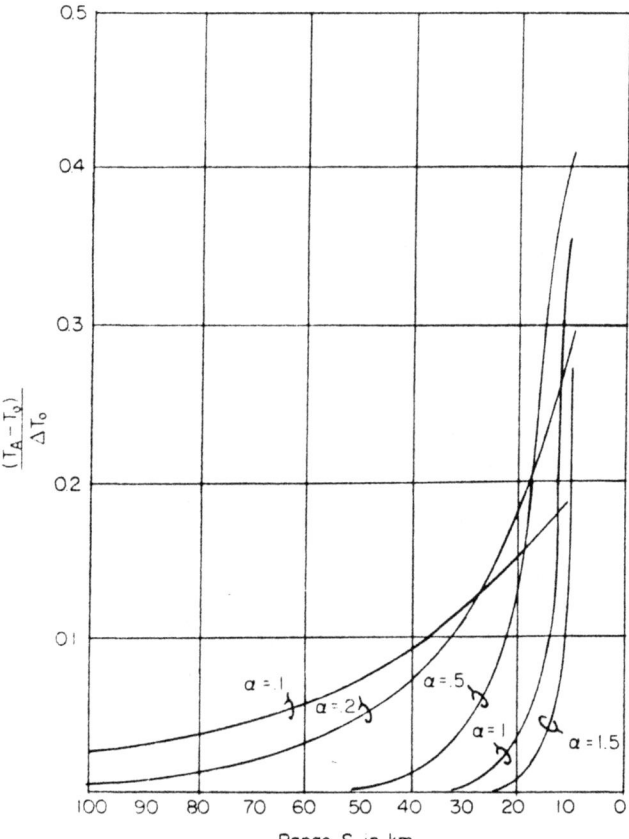

F<small>IG</small>. 26. The normalized multifrequency response of a radiometric sensor to a ramp temperature anomaly of ΔT_0, 10 km in extent, at a horizontal range S from the sensor. See Fig. 25 for comparison of the response characteristics for various values of the atmospheric absorption coefficient α.

of observation.

A dual channel 5 mm radiometer sensor was developed in 1968 by the Propagations Studies Branch of the NASA Electronics Research Center to investigate CAT detection capability. A photograph of this millimeter radiometric sensor is shown in Fig. 28. The antenna is a single conical lens fed horn. The antenna output is fed via an orthogonal mode transducer to the inputs of two radiometric receivers. One receiver is tunable over the frequency range from 51 to 53 GHz and the other from 57 to 59 GHz.

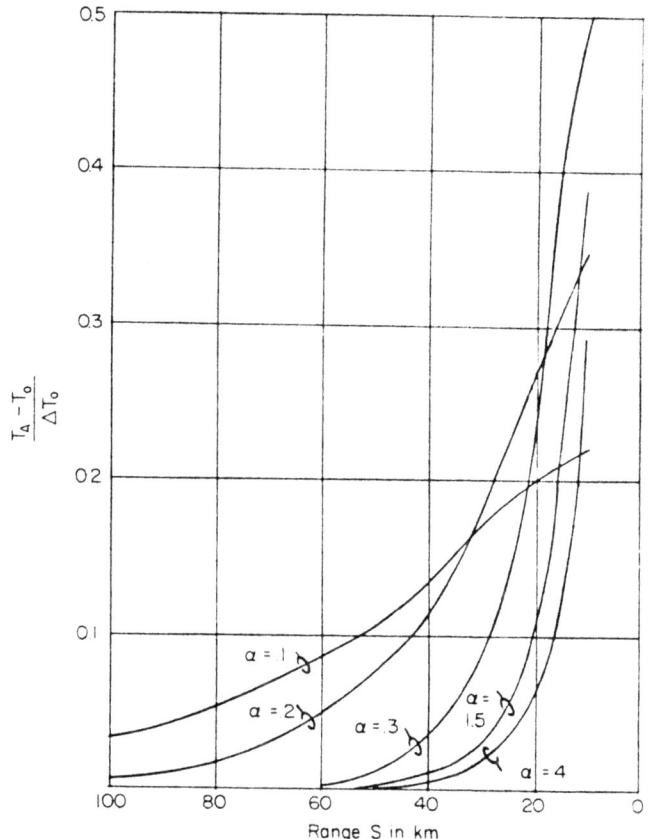

FIG. 27. The normalized multifrequency response of a radiometric sensor to an
exponential temperature anomaly which has an average value of ΔT_0
over a range interval 10 km in extent, at a horizontal range S from the
sensor. (See Fig. 25 for comparison of the response characteristics for
various values of the atmospheric absorption coefficient α.)

The individual tuning ranges and the frequency separation between channels
allow selection of corresponding atmospheric absorption coefficient values
for observation of 0.1 and 1 dB/km for flight altitudes from 30,000 to
60,000 feet.

The first aircraft flight test has been scheduled for the fall of 1969.
As a part of this flight program, it is planned to point the antenna verti-
cally down from high altitude. This should provide an interesting evalu-
ation of vertical sounding of the atmospheric temperature profile in the

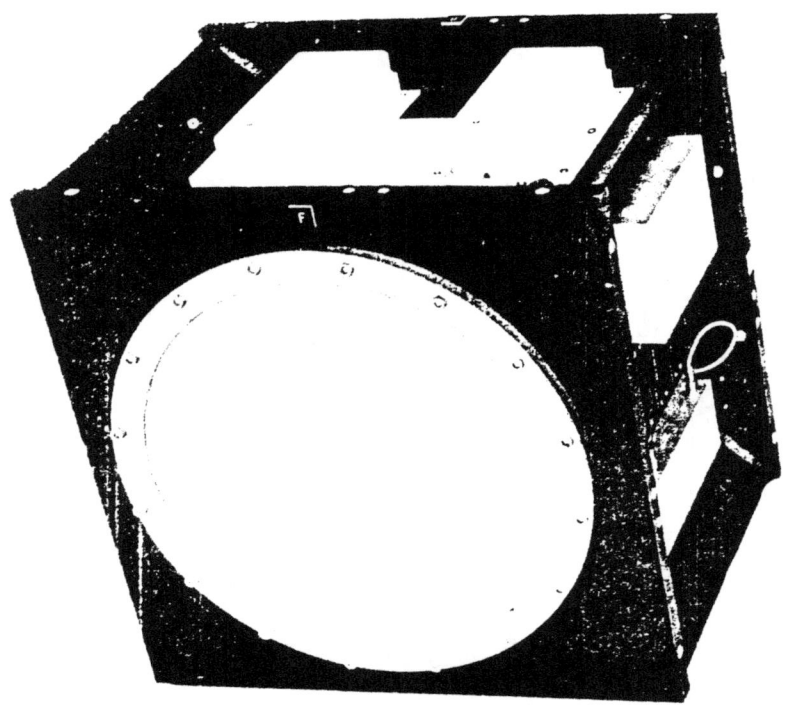

FIG. 28. 5-mm wavelength dual channel radiometric sensor developed by the NASA Electronics Research Center to experimentally verify the ability to detect clear air turbulence regions along the forward flight path of high altitude aircraft.

lower troposphere. The diversity of the two disciplines, CAT detection and meteorological measurements, will be joined by one sensor as a common denominator in this dual experiment to explore and exploit.

IV. A LOOK INTO THE FUTURE

Historically, our communication needs and associated technological requirements have provided the stimulus for expanding our radio capabilities from long to shorter wavelengths. There has been no slackening of the pace as this need now focuses attention on the 3 cm to 3 mm wavelength region. This time, however, the communicator has a silent and persistent partner already actively exploring this region of the spectrum with definite plans for exploitation. The ability to obtain a global picture of atmospheric water vapor and temperature distributions, combined with air mass circulation under clear air conditions, offers the potential to predict in advance the formation of storm clouds and their motions. Several significant applications can be accomplished only in this portion of the spectrum as a consequence of the nature of the physical processes involved.

In the future, as communicators look through the atmospheric windows, microwave meteorologists will be measuring the global structure of the atmosphere in the spectrum between the windows. They may both join together, however, in earth orbit at operating wavelengths in the vicinity of 5 mm, since here the communicator is assured an environment free of man-made electromagnetic interference, protected by the several hundred dB attenuation of the oxygen blanket surrounding the earth. Here, the needs of the meteorologists will parallel those of the present day radio astronomer in reaching agreements on "quiet bands" to be used exclusively for passive remote sensing. This is not an insignificant problem. Without careful consideration, harmful interference to passive studies and applications may result. For example, a microwave mapping radiometer has been proposed for an experimental program on the Nimbus series of satellites. A frequency of 19.35 MHz was chosen because it is in the region of the spectrum where the brightness temperature of smooth sea water is practically independent of the water temperature. This frequency is, coincidently, in a radio astronomy band presently protected from man-made electromagnetitc transmission. It was recently suggested that this particular radio astronomy band be relocated to 23.55 GHz in order to make way for space-to-earth communications. The advent of space-to-earth communication systems at selected frequencies of this type may seriously effect remote sensing applications which cannot change frequency.

The explosive exploration of the 3 cm to 3 mm wavelength region will continue at an accelerating pace since instrument capability is no longer the limiting factor. Within the next half decade a radiometric temperature sensing capability of better than 1°K will be achievable throughout this entire wavelength region, with postdetection integration time constants no greater than 1 sec. The challenge will be to extract knowledge and understanding from the centimeter to millimeter wavelength signals that are naturally emitted by the atmosphere, the oceans, and all surface terrain materials. It is never an easy task, however, when the unknown is so close to home. As J. P. Wild said at the Fourth Pawsey Memorial Lecture at the University of Queensland, Australia, in April of 1968, speaking about our knowledge of the sun: "You see the sun is rather an enigma in astrophysics. We appear to know so much about astrophysics—about the galaxy and the universe and so on; there might be a few controversial alternatives when astronomers talk of cosmology or quasars or pulsars, but on the whole they sit back and review their achievements with remarkable satisfaction. . . . When you know two or three things about something there is no difficulty in producing a theory to explain it. But when you know a thousand things, the theory becomes more difficult, and when you know that another thousand things are waiting to be discovered the theorists get frightened off. And so, apart from a few of the braver theorists, the onus is left in the hands of experimental physicists: especially those prepared to persevere gradually, step by step, with the scientific method; and especially those prepared to fashion new lines of attack."

As we enter this new era with the capability to passively and remotely sense the location, identity, and condition of our earth resources from satellite orbit, our success will be measured by our perseverance and integrity to exploit the microwave spectrum in the best interest of all mankind.

REFERENCES

1. Dicke, R. H. The measurement of thermal radiation at microwave frequencies. *Rev. Sci. Instr.* **17**, 268-275 (1946).
2. Haroules, G. G., Brown, W. E., III, and Ewen, H. I. Method and Means for Providing an Absolute Power Measurement Capability. Patent application, February, 1967.
3. Dicke, R. H., Peebles, P. J. E., Roll, P. G., and Wilkinson, D. T. Cosmic black-body radiation. *Astrophys. J.* **142**, 414-419 (1965).
4. Thompson, W. I., III, and Haroules, G. G. A review of radiometric measurements of atmospheric attenuation at wavelengths from 75 centimeters to 2 millimeters. NASA TN-D-5087 (January 1969).
5. Cheung, A. C., Rank, D. M., Townes, C. H., Thornton, D. D., and Welch, W. J. Detection of NH_3 molecules in the interstellar medium by their microwave emission. *Phys. Rev.* **21**, 1701-1705 (1968).
6. Van Vleck, J. H. Absorption of microwaves by water vapor. *Phys. Rev.* **71**, 425 (1947).
7. Becker, G. E., and Autler, S. H. Water vapor absorption of electromagnetic radiation in the centimeter wave-length range. *Phys. Rev.* **70**, 300 (1946).
8. Ho, W., Kaufman, I. A., and Thaddeus, P. Laboratory measurement of microwave absorption in models of the atmosphere of Venus. *J. Geophys. Research* **71**, 5091 (1966).
9. Straiton, A. W., and Tolbert, C. W. Anomalies in the absorption of radio waves by atmospheric gases. *Proc. IEEE.* **48**, 898-903 (1960).
10. Barrett, A. H., and Chung, V. K. A method for the determination of high altitude water-vapor abundance from ground-based microwave observations. *J. Geophys. Research* **67**, 4259 (1962).
11. Artman, J. O., and Gordon, J. P. Absorption of microwaves by oxygen in the millimeter wavelength region. *Phys. Rev.* **96**, 1237 (1954).

12. Anderson, R. S., Smith, W. V., and Gordy, W. Microwave spectrum of oxygen. *Phys. Rev.* **87**, 571 (1952).

13. Meeks, M. L., and Lilley, A. E. The microwave spectrum of oxygen in the earth's atmosphere. *J. Geophys. Research* **68**, 1683 (1963).

14. Westwater, E. R., and Strand, O. N. Application of statistical estimation techniques to ground-based passive probing of the tropospheric temperature structure. U. S. Dep't. of Commerce, ESSA Technical Report IER 37-ITSA 37, Boulder. Colorado (1967).

15. Gora, E. K., The rotational spectrum of ozone. *J. Mol. Spectroscopy* **3**, 78 (1959).

16. Van Vleck, J. H. The absorption of microwaves by oxygen. *Phys. Rev.* **71**, 413-424 (1947).

17. Gautier, D. and Robert, A. Calcul du coefficient d'absorption des ondes millimetriques dans l'oxygene moleculaire en presence d'un champ magnetique faible. application a l'atmosphere terrestre. *Ann. Geophys.* **20**, 480 (1964).

18. Lenoir, W. B. Microwave spectrum of molecular oxygen in the mesosphere. *J. Geophys. Research* **73**, 361 (1968).

19. Stafford, L. F., and Tolbert, C. W. Shapes of oxygen absorption lines in the microwave frequency region. *J. Geophys. Research* **68**, 3431-3435 (1963).

20. Reber, E. E., Mitchell, R. L., and Carter, C. J. Oxygen absorption in the earth's atmosphere. Air Force Report No. SAMSO-TR-68-488, Aerospace Report No. TR-0200 (4230-46)-3 (1968).

21. Van Vleck, J. H., and Weisskopf, V. F. On the shape of collision broadened lines. *Revs. Modern Phys.* **17**, 227-236 (1945).

22. Staelin, D. H. Measurements and interpretation of the microwave spectrum of the terrestrial atmosphere near 1-centimeter wavelength. *J. Geophys. Research* **71**, 2875-2881 (1966).

23. Gaut, N. R. Studies of Atmospheric Water Vapor by Means of Passive Microwave Techniques. Ph. D. Thesis, Dept. of Meteorology, Massachusetts Institute of Technology (1967).

24. Trambarulo, R., Ghosh, S. N., Burrus, C. A., Jr., and Gordy, W. The molecular structure, dipole moment, and a *g* factor of ozone from its microwave spectrum. *J. Chem. Phys.* **21**, 851-854 (1953).

25. Hughes, R. H. Structure of ozone from the microwave spectrum between 9,000 and 45,000 Mc. *J. Chem. Phys.* **24**, 131-138 (1956).

26. Walshaw, C. D. Line widths in the 9.6 μ band of ozone. *Proc. Phys. Soc. London* **A68**, 530 (1955).

27. Mouw, R. B., and Silver, S. Solar radiation and atmospheric absorption for the ozone line at 8.3 mm. *Inst. Eng. Res. Ser.* 60/277, University of California, Berkeley (1960).

28. Caton, W. M., Welch, W. J., and Silver, S. Absorption and emission in the 8-mm region by ozone in the upper atmosphere. *Space Sci. Lab.*, Ser. No. 8. Issue 42 (1967).

29. Barrett, A. H., Neal, R. W., Staelin, D. H., and Weigand, R. M. Radiometric detection of atmospheric ozone. *Quart. Prog. Rept., Res. Lab. of Electronics*, M. I. T. (July, 1967).

30. Caton, W. M., Mannella, G. G., Kalaghan, P. M., Barrington, A. E., and Ewen, H. I. Radio measurement of the atmospheric ozone transition at 101.7 GHz. *Astrophys. J. (Letters)* **151**, L 153 (1968).

31. Caton, W. M., Private communication

32. Hunt, B. G. Photochemistry of ozone in a moist atmosphere. *J. Geophys. Research* **71**, 1385 (1966).

33. Radiometric characteristics of the atmosphere for reference-direction sensing in space vehicle navigation. Air Force Contract AF 19 (628) 3239.

www.ingramcontent.com/pod-product-compliance
Lightning Source LLC
Chambersburg PA
CBHW051222170526
45166CB00005B/2010